PERILOUS BOUNTY

PERILOUS BOUNTY

The Looming Collapse of American Farming and How We Can Prevent It

TOM PHILPOTT

BLOOMSBURY PUBLISHING
NEW YORK • LONDON • OXFORD • NEW DELHI • SYDNEY

BLOOMSBURY PUBLISHING
Bloomsbury Publishing Inc.
1385 Broadway, New York, NY 10018, USA

BLOOMSBURY, BLOOMSBURY PUBLISHING, and the Diana logo are trademarks of Bloomsbury Publishing Plc

First published in the United States 2020

Graphs courtesy Gary Antonetti/Ortelius Design

ISBN: HB: 978-1-63557-313-8; eBook: 978-1-63557-314-5

LIBRARY OF CONGRESS CATALOGING-IN-PUBLICATION DATA IS AVAILABLE

Library of Congress Control Number: 2019012345

2 4 6 8 10 9 7 5 3 1

Typeset by Westchester Publishing Services
Printed and bound in the U.S.A. by Berryville Graphics Inc., Berryville, Virginia

For Alice Brooke Wilson, always and forever;
for the workers who feed us; and
for Anne Sugrue (1941–2019)

CONTENTS

Introduction

Salad greens love an autumn chill. It concentrates their flavor, making them peppery but also sweet. And cold weather beats back the insect pests that besiege these tender leaves in warm months.

To cut delicate greens on frosty mornings, you can wear gloves, but for dexterity's sake, they have to be the kind that leaves the fingers bare. Each day when you start out, your fingers take on a stinging chill. We had a joke at the small organic farm in the Appalachian Mountains of western North Carolina where I worked in the early 2000s: when the pain has finally eased because your fingers have gone completely numb, you know you're halfway through the morning's harvest.

The daily ache was worth it. The farm cultivated broccoli, chard, beets, potatoes, squash, green beans, and more; but fall salad greens were the crown jewel. Spinach, young kale, arugula, and a variety of spicy mustards were its lifeblood, its claim to fame. You don't have to take my word for it. After a visit in 2005, the food writer Jim Leff, founder of the website Chowhound, called them "hallucinogenic in their intensity and persistence of flavor; coated with a dab of oil and vinegar, they steal every meal they accompany."

My years on the farm were a crash course in some of the challenges and paradoxes facing U.S. agriculture, and those prized salad greens figured in one memorable lesson.

One chilly afternoon, late in fall 2006, I was in the farmhouse calling restaurant chefs to take their delivery orders, a key part of our business. I reached the chef of a much-loved local restaurant, a reliable customer. On our call, he ordered his usual amounts of other fall goods: garlic, beets, mature chard. The conversation turned to salad greens, which he had been buying from us in large quantities for years. This time, he stopped short: "I hate to ask, but can you go down on price?" He explained that his food distributor had just begun offering organic salad greens shipped from California. "They're pretty good quality. Not nearly as good as yours, but not bad either—and they're less than half the price."

I thought for a second. I didn't feel like telling him what trouble it was to cut the greens he had been buying, or that we relied on the price we had established to get through the winter and buy the spring's seeds. I simply explained that fall greens were our central crop, and we couldn't go down on price. He bought a few bags—a fraction of his usual order, for use on the nightly special, he said.

Not long after, such haggling became the norm among most of our restaurant customers for nearly everything we grew. I didn't know it at the time, but that chef was gently informing me that "not bad" organic vegetables grown in California had emerged as the price setter for produce among our area's farms.

Of course, at that time, California fruits and vegetables already dominated our area's supermarkets, as they did supermarkets throughout the country. For midsize, nonorganic farms across the nation, the "California price" has long served as a price ceiling. The downward pressure of that price, in turn, made each acre of farmland less profitable, forcing growers to strive to get bigger or exit the business. But the California agricultural behemoth was even big-footing its way into the relationship between a niche organic mountain farm and its restaurant market. The long-fomenting revolt against industrially produced food was beginning to consume its own.

That same year, Michael Pollan published his landmark *The Omnivore's Dilemma*, which further galvanized awareness of our food system's depredations. Sales of organic food were booming; farmers markets were multiplying across the country. Even amid this surge in desire to "know where your food comes from," California's massive farms were intensifying their grip on U.S. food production. California organic was squeezing local organic.

According to conventional economics, our plight reflected the invisible hand of the market slapping down inefficient producers in favor of ones who can do a better job of growing vegetables. The principle of comparative advantage, which has been with us since the nineteenth century, holds that each region should generate a surplus of stuff it can produce most cheaply and trade those excess goods for the specialized products of other regions. "The case for specialization is perhaps nowhere stronger than in agriculture, where the costs of production depend on natural resource endowments, such as temperature, rainfall, and sunlight, as well as soil quality, pest infestations, and land costs," the Duke University economist Steven Sexton argued in a 2011 piece for the *Freakonomics* website.

Sexton went on to cite California's farm regions, with their "mild winters, warm summers, and fertile soils," as a prime example of comparative advantage in action. Sexton pointed approvingly to the state's dominance of the American supermarket produce aisle and warned consumers against supporting local efforts to challenge it, which would only "raise the cost of food by constraining the efficient allocation of resources."

Sexton couldn't have known it then, but that very year, California had entered what would turn out to be its most severe drought in at least five hundred years. In a sense, the succession of dry years vindicated the claim that California should be our dominant vegetable and fruit producer. Through its long, parched stretch, the California agriculture juggernaut lurched forward. The long drought caused dry wells, poisoned water, and other massive hardships for the people who toil in the state's blazing-hot agricultural valleys. But they caused barely a blip for most U.S. consumers, who encountered the same reliable bounty in their supermarkets. In 2010, the year before the epic drought, California farms churned out $20.7 billion worth of fruits, vegetables, and nuts, about 53 percent of the U.S. total. In

2015, at the drought's height, output had actually risen to $28.6 billion, accounting for 57 percent of all such U.S. output.

For most Americans, it's easy to ignore the Central Valley, even though it's as important to eaters as Hollywood is to moviegoers or Silicon Valley is to smartphone users. Occupying less than 1 percent of U.S. farmland, the Central Valley churns out a quarter of the nation's food, predominantly fruit, vegetables, and nuts.

In this book, I'll show how this apparent triumph of comparative advantage is like the sight of water in a sunbaked desert: a mirage, and a dangerous one for U.S. eaters.

While California is the undisputed top supplier of our fruits and vegetables, another region, the Midwest, occupies a similar position for the foodstuff that takes the central position on the U.S. dinner plate: meat. The Midwest's store of rich, black, prairie-derived topsoil places the region among the globe's great breadbaskets. Its farms supply the bulk of corn and soybeans that provide the feed for the United States' highly industrialized meat sector. Just as you'll find loads of California vegetables in Des Moines all summer, meat cases from San Diego to San Francisco are stuffed with cuts from animals fattened on Iowa corn. The corn and soybeans grown there also course through the processed- and fast-food industries, providing the bulk of the fat and much of the sweetener that makes prefab food palatable.

The Corn Belt system, too, enjoys an apparent comparative advantage that generates an endless bounty of cheap meat even as it exerts downward pressure on small farms nationwide that reject the industrial model. Again, the comparative advantage on display is largely spectral.

In short, this is a book about the vast swaths of fertile land that Americans rely on for sustenance but that relatively few ever see or smell. The premise is simple: the United States has two dominant food-producing regions—California's Central Valley and the former prairielands of the Midwest—and both are in a state of palpable and accelerating ecological decline. At the moment, the effects are mostly felt by the workers who make the farms, groves, and feedlots hum, and who are subjected to increasingly tough conditions; in addition to the baseline rigors of their jobs, they endure fouled water, putrid air, and the decay of public services that accompany a declining population. But while their plight is easy for many Americans to ignore, there's something else afoot in these regions that will affect every U.S. resident who eats: to grow our food, I argue in this book, the agribusiness interests that dominate the Central Valley and the Corn Belt are also actively consuming the ecological foundations that support agriculture itself.

We start our journey in California. The 2011–17 drought, compounded by anomalously hot weather, made national headlines. It ended only when drowned by California's wettest winter in a century. Yet California's water crisis is ongoing, and worsening, independent of short-term drought cycles. Essentially, the farms we rely on to stock the supermarket produce shelf have gotten so ravenous for irrigation that, even accounting for wet years, they steadily overdraw California's rather modest water resources. These farms drain aquifers and place increasing demands on the once-mighty rivers that originate in the Sierra Nevada range. Climate change tightens the state's water squeeze; it's one of the engines driving the ever-hotter heat waves that force farmers to irrigate more, and also the ever-more-frequent droughts that compel them to tap millennia-old underground aquifers.

But drought isn't the only threat that looms over the U.S. fruit-and-vegetable patch. In California's chaotic weather regime, severely dry stretches are intimately linked with their opposite: cataclysmic deluges. The Great Flood of 1861–62 buried the entire Central Valley under ten feet of water just after California achieved statehood in 1860, transforming the region's agriculture. The event, by far the most violent flood in California's postconquest history, has largely vanished from popular memory. But a growing body of research suggests that such megastorms occur every one hundred to two hundred years—and a warming climate makes one of equal or greater magnitude probable within the next decades.

Next, we move 1,700 miles due east, where corn and soybean farms exist at a similar intersection of massive production and festering crisis. The Corn Belt's output makes our factory-scale cow, hog, and chicken farms hum, feeding our meat habit (and increasingly, that of China and other nations). The Midwest's densely planted fields cover a combined landmass one and a half times the size of the entire state of California, dominated by just two crops.

Whereas California's farms are essentially mining nonrenewable water, Corn Belt operations are quietly exhausting an even more tenuous resource: soil. When U.S. settlers arrived in the nineteenth century, up to sixteen inches of fertile loam blanketed the prairie, the result of hundreds of thousands of years of interaction between bison and the grasses that sustained them—a process managed by Native Americans for several millennia before the United States enclosed the region. Since the plowing of the prairie, at least half that topsoil has leached away.

After the 1930s Dust Bowl crisis, the U.S. Department of Agriculture put into place conservation policies that for a while slowed the rate of soil erosion. But with the great corn and soybean boom that ignited in the mid-2000s, the curse of erosion has roared back. According to Iowa State University researchers, soil in Iowa—the Corn Belt's linchpin—is eroding around sixteen times faster than the natural soil-replacement rate. As the soil washes away, it undermines the very basis of production in our breadbasket. So, too, do the vast amounts of chemicals farmers apply, poisoning water in cities like Des Moines, Columbus, and Toledo, and all the way downstream to the Gulf of Mexico, where fertilizer runoff from farms generates an annual oceanic dead zone the size of New Jersey.

As with California's water woes, climate change speeds up the process. The warming atmosphere brings weather chaos; between 2003 and 2017, central Iowa endured no fewer than four "hundred year" storms. When heavy rains pummel bare fields in the spring, long gashes form in the land, rapidly stripping away tons of soil and dispersing it (and plenty of freshly applied fertilizers and pesticides) into streams and rivers. Farmers typically respond to these increasingly frequent events by flattening out their fields—pushing fresh dirt into the gaps and planting it as usual the next year—thus loading them up for the next major storm. Agriculture in the region has emerged as a machine for sacrificing soil, at a time when farms need to be building soil to prepare for coming weather shocks.

Again, like California's hot, rugged agricultural zones, the Corn Belt's farms operate in the shadows of our food culture. For those who do dare to look, the response is often to opt out—to "eat local," to shop at the farmers market. While that choice is perfectly rational, and has led to an impressive boom in local and regional food sales since the mid-1990s, it has done little to slow the intensification or ecological degradation of industrial agriculture. Indeed, as I found on that Appalachian farm in the 2000s, output from California and the Corn Belt imposes cutthroat price competition on smaller alternatives, limiting their profitability and confining them to niche status. After two decades of rapid growth in farmers markets and other farm-to-table efforts, the massive farms of those two regions still supply the vast majority of food Americans eat. If the current ecological unraveling proceeds apace, it will trigger social and ecological disruptions that "voting with your fork" won't protect you from.

In this book, I explore how the bounty on offer in our supermarkets and restaurants consumes the very resources on which it relies. When most Americans follow the standard dietary advice to eat more fruits and vegetables, they're essentially taking a bite out of California—or more precisely, a big swig of its increasingly scarce water supply. And when we "go Paleo" and follow current dietary trends by amping up our protein intake, tucking into a steak, a pile of bacon, or a chicken breast, we can thank the former prairielands of the U.S. Midwest and its vanishing store of topsoil.

What drives this creeping disaster is the rise of a virtual oligarchy of companies that capture most of the profit generated by the trillion-dollar-a-year food economy. Three massive, globe-spanning companies—Bayer-Monsanto, Corteva, and Syngenta—sell the great bulk of the seeds and pesticides available to U.S. farmers. A handful of others—Tyson Foods, Cargill, JBS, and Smithfield Foods—slaughter and pack the majority of meat we eat. The market for trading corn and soybeans largely belongs to Cargill and Archer Daniels Midland. In California, a single firm—the privately held Wonderful Company—dominates the water-sucking almond, pistachio, pomegranate, and mandarin-orange markets. As I'll show, these behemoths profit by squeezing farmers and offloading the costs of the ecological degradation they cause onto communities and taxpayers. And they invest a serious portion of their gains into Washington, D.C., lobbying and campaign donations—allowing them to beat back regulations and resist challenges to their primacy in the food system.

Throughout the book, I focus on the handful of seed-pesticide corporations, investment funds, and magnates who benefit from these dire trends, with on-the-ground dispatches featuring scientists documenting the damage and farmers who are pushing back. But the story does not inevitably point to a looming doomsday of scarcity and hunger. Rather, in the book's final chapters, I profile a still-tiny band of farmers who are already showing the way toward an abundant future.

A brief note on what this book is and isn't. *Perilous Bounty* offers a big-picture view of what corporate-dominated industrial agriculture is doing to our land and water resources and what it means for our food supply as we plunge into an era of climate chaos. This book isn't a comprehensive look at the entire geography of U.S. agricultural hotspots. Other regions engage in industrial-scale meat production: eastern North Carolina houses an extraordinary concentration of massive hog facilities, and factory-like chicken barns dot landscapes from the Chesapeake Bay through the Deep South to Arkansas. Fruits and vegetables get pumped out at high volume in Central Florida (see Barry Estabrook's great 2011 book *Tomatoland: How Modern Industrial Agriculture Destroyed Our Most Alluring Fruit*) and in Texas' Rio Grande Valley. These places play a vital role in providing our food, and they're undergoing their own ecological and social crises. I focus on California and the Upper Midwest because they are by far the biggest linchpins of our sustenance.

There's also the crucial question of what industrial agriculture does to the workers it relies on and to the people who live within its proximity. Toxic water, fouled air, low wages, horrible working conditions, the withering away of public services—these facts of life in our industrial agriculture zones are signals of a food system gone rancid. With a few exceptions, like the annual agrichemical-fed algae bloom that blots out life in the Gulf of Mexico, their harms tend to fall most heavily on the people who live nearby, allowing most Americans the privilege of enjoying burgers and salad without thinking about, say, entire towns that are forced to buy bottled water to avoid being poisoned. But people in the Central Valley and the Corn Belt aren't quietly enduring their lot. Groups including the Central California Environmental Justice Network, the Iowa Citizens for Community Improvement, and others are organizing communities to demand the right to clean air and water encoded in U.S. law. These efforts receive little national attention, but they have the potential to benefit everyone, because ensuring clean water for residents of Kern County, California, or Hardin County, Iowa, would mean a switch to farming practices that respect soil and water resources. Properly chronicling these efforts would require an additional book. For this one, I focused on the way industrial agriculture, as currently practiced, threatens the food security of everyone who now relies on it for sustenance.

Then there's the question of land ownership. U.S. settlers violently seized the lands this book focuses on from Native Americans who had lived on them for generations. It's beyond the scope of this book to delve into the horror of that theft or its enduring aftermath of oppression; but I do detail the systematic abuse that U.S. control has visited upon these incredibly fertile and productive ecosystems, which had flourished over millennia under stewardship by indigenous populations and are now on the verge of collapse. While chattel slavery never prevailed in California or the Corn Belt, enslaved people provided the labor that fueled the U.S. economy during its great nineteenth-century expansion westward. These abuses are part of the festering debt owed by U.S. white-settler society to American Indians, Mexican Americans, and Black Americans today, and I hope my book can aid in the movement for reparations.

I also hope that *Perilous Bounty* can serve as a weapon for the environmental- and worker-justice movements afoot in California farm country and the Corn Belt, because the struggle for decent living conditions in farm country is also a struggle for a food system that can feed everyone for generations.

1

High and Dry

On a sunny September morning amid the dusty fields of the San Joaquin Valley's west side, Joe Del Bosque slowly guided his late-model white GMC truck down a dirt road that separates two of the main fields of his two-thousand-acre farm. The field to our right featured endless rows of low-slung, creeping melon plants; to the left, there was a thick grove of almond trees, their branches heavy with nuts ready for harvest.

The GMC's vanity license plate advertised which of the two crops is his main stock in trade, the lifeblood that has sustained his career: MELONS. Growing them has been the family profession for three generations. In the early 1900s, Del Bosque's grandfather migrated from Mexico to work melon fields in the Imperial Valley, a desert farming region on California's southern border irrigated with water from the Colorado River. By the 1950s, when federally funded irrigation canals brought steady water to the parched region west of Firebaugh, growers began devoting acres to melons there, too. Del Bosque spent his childhood migrating between the two regions, following the melon harvest, picking fruit alongside his parents. His father eventually established himself as foreman of a large melon farm. After earning a degree in agriculture from nearby California State University, Fresno, Del Bosque went to work for the same operation as his dad, who taught him how to manage the production of melons on a grand scale.

Then came the economic recession of the early 1980s: a period marked by low crop prices, collapsing land values, and a spike in farm failures. For young Joe, the crisis created an opportunity, and he hustled his way to a foothold as a landowner, starting out as essentially a sharecropper, selling melons he grew with borrowed money on rented land. Today, Del Bosque grows eight hundred acres of his prized crop, making him one of the nation's biggest producers of organic melons—on the very fields he once toiled as a child melon picker. His produce is sold as far away as Florida through his biggest customer, Whole Foods. He acknowledges that he's a walking statistical anomaly in this region of California: a farmworker kid who grew up to be a major grower, employing hundreds of farmworkers in turn.

To our right, a crew of those workers was harvesting under a fast-warming September-morning sun. Six men walked briskly behind a platform pulled by a tractor, tossing ripe melons to three women on board, who packed them into boxes, occasionally tossing out a fruit judged to be flawed.

Del Bosque's success against heavy odds required optimism. These days, though, he is worried. Both of his crops—melons and almonds—are in high demand. But both face specific challenges that make their futures uncertain. In this sense, Del Bosque Farms is a microcosm of the threatened San Joaquin Valley—and the American agricultural system as a whole. It's as good a place as any to begin this story.

A kindly sixty-something in a blue-striped button-down shirt, jeans, and a white cowboy hat, Del Bosque explained to me that his melon harvest is extremely labor-intensive. During the season, which stretches from late June into October, he has nearly three hundred people working his fields. He favors old varieties that are prized for their flavor and resilience to attacks from pests, making them great for organic farming. But their delicate fruits ripen at haphazard intervals, requiring several passes over two weeks on each section, by crews of skilled workers who can quickly determine which orbs to snap up and which need more time. Most melon growers plant newer varieties that ripen uniformly and are hardy enough to be picked by machines. Their efficiency sacrifices flavor, he says—it's the reason those melon chunks you find in places like hotel breakfast buffets taste like nothing.

The market for organic fruit is growing fast. When I visited him in 2018, Whole Foods had bought more of his melons than ever. But Del Bosque believes his melon empire's days are numbered; he'll probably have to phase them out before he passes his farm business on to his adult children and their spouses. The issue, he said, is labor costs.

There are two labor-related factors at play that will likely push Del Bosque Farms to abandon the crop on which it has been built. The first is the ramped-up federal enforcement of immigration law. Labor shortages started affecting Del Bosque around 2010, under President Barack Obama. In its first term, the Obama administration initiated a deportation frenzy to gain credibility with Republicans, in an ultimately failed effort to pass comprehensive immigration reform. The labor situation "just slowly got worse" under Obama, Del Bosque said.

Then Donald Trump took the presidency. "The difference is, this guy's loud," Del Bosque explained. Since the 2016 election, and amid the president's recurring rhetorical battles with California's Democratic leadership, the state's farmworkers "have gotten more nervous," he said. "At the beginning of the season, a lot of folks were reluctant to leave home" and head to the fields to work the 2018 harvest, for fear of being pulled over and ordered to show papers. Eventually, Del Bosque—who also runs a farm-labor contracting business with his wife—cobbled together a workforce sufficient to profitably bring in the harvest. But the anxiety among his workers was palpable, and "lots of folks are leaving the area and going to places that are less under the microscope," like Oregon. Melons are a "finicky crop," Del Bosque said, and it was "getting harder and harder to find workers."

At that point, we left the car and ventured into the field, trailing behind the work crew as they snagged melons. Del Bosque stooped to pluck a green-gray cantaloupe from the ground amid a tangle of vines, held it in his hand, judged it ripe, produced a knife from his jeans, and sliced into it. A warm, sweet aroma rose. He offered me a slice, and I bit in, finding a delicate texture and a burst of flavor. Warm from the sun and juicy, it was pure melon deliciousness: round, sweet notes balanced by just enough acidity.

As the workers steadily made their way up the field, we headed back to the truck, and Del Bosque told me about the second labor-related factor that haunts his business: these workers were due for a substantial pay raise.

The federal government established the forty-hour workweek, eight-hour day, and standards for overtime pay in 1938 during the New Deal era. At the behest of Southern Democrats determined to maintain the power to exploit the region's largely black farm labor force, the law excluded agricultural labor. In a 2016 vote, the California legislature became the first statehouse in the nation to extend the same overtime benefits to farmworkers that apply to other workers.

By 2022, the state's farm operations have to start paying workers time and a half for every minute worked over eight hours per day and forty hours per week. Meanwhile, by 2023, the state's minimum wage will rise to fifteen dollars per hour—up from eleven dollars in 2018—and continue to increase at the rate of inflation thereafter. Since its first stirrings after the gold rush in the nineteenth century, California's farming behemoth has run on cheap labor. Starting very soon, it will be forced to pay its workers under the same rules as other industries.

Del Bosque said the new regime will make it impossible to profitably churn out tasty melons. During the harvest, he needs crews in the field seven days per week, often for more than eight hours a day, or prime fruit rots unharvested. "Yesterday was a big day—we had crews out from seven A.M. until six P.M." On a day like that, under the longstanding rules, overtime only kicks in after ten straight hours, entitling workers to one and a half times their wage for the single additional hour. The new rules will mean three hours of overtime for such an eleven-hour day—and the simultaneously rising minimum wage means a base pay rate about 35 percent higher than the current rate.

In Del Bosque's view, a future marked by labor scarcity and elevated wages would be inhospitable to labor-intensive crops like melons. The day his kids will take over the farm isn't too far off. "I don't know if they will be able to keep the melons going," he said. "I don't know about the future of melons here in California." He speculated that other growers would continue with the fruit, but they'd concentrate on varieties adapted for mechanical harvesting, with a flavor and texture resembling the cardboard box they're packed in.

Already, other crops that require lots of hand labor, including asparagus, were being phased out of the region and moving to Mexico, where the prevailing farm wage is eight dollars per day. High labor costs and competition with Mexican farms forced Del Bosque to cut his asparagus patch from 200 acres to 100 acres. Statewide, the same forces drove asparagus cultivation from 22,500 acres in 2006 to 8,000 in 2018.

Our conversation turned to a part of the operation that had expanded in recent years: the almond trees to our left. They had fast emerged as the rival to melons as Del Bosque's main crop. But his almond business, too, was imperiled, although not by labor trouble. One reason almond acreage doubled in the San Joaquin Valley between 2000 and 2018 is that almonds require essentially no hand labor: planting, irrigation, and harvesting are all done by machines. What troubled Del Bosque's 625-acre plantation of almond trees—and made the prospect of expanding it anxiety-inducing—was the increasingly uncertain access to irrigation water.

Almond trees originated in the Middle East and Central Asia, and have flourished at least since biblical times in the Mediterranean zones of southern Europe. They need precisely the hot, dry summers and mild winters that characterize the San Joaquin Valley. But when densely planted over vast expanses, they require titanic amounts of irrigation water. It takes about a gallon of water to grow a single almond, and California's one million acres of groves—covering about a fifth of the farmland in the San Joaquin Valley, and still expanding fast—now consume four times more water annually than the city of Los Angeles. Del Bosque said that one factor that keeps him clinging to melons despite the labor crunch is that melons require about half the water per acre as almonds.

He farms in what's essentially a desert, with an average annual rainfall of nine inches. (Iowa, for comparison, gets upwards of thirty-five inches;

deserts are generally considered places that receive less than ten inches of rainfall annually.) Summers are blazing hot, with average daily temperatures in July hitting 97 degrees Fahrenheit. They're also bone-dry—nearly all the rain that does fall occurs during the mild winters.

To make his farm bloom with so little water from the sky, Del Bosque and his San Joaquin Valley peers rely on water from two sources. The first is the annual snowmelt from the Sierra Nevada to the east, shunted through a network of dams, aqueducts, and canals. (That jagged, majestic, snow-topped mountain range is essentially the garden tap for your supermarket's cornucopia.) The second is wells tapped into aquifers beneath the valley's floor.

The two sources are intimately related: every spring, water gushes down those mountains, accumulating over millennia, and filling up those underground water reserves. And when a given year's snowmelt is paltry, farmers revert to the pump, drawing down their aquifers. When snowmelt is plentiful, they ease up on their wells, and aquifers can (at least theoretically) recharge. Both sources are in a state of crisis.

San Joaquin Valley farmers, including Joe Del Bosque himself, tend to vote Republican and dismiss climate science. But they've observed remarkable changes over the course of their careers. The average amount of snow annually captured by the Sierra Nevada declined by as much as 20 percent between the 1980s and 2000s. A group of researchers with Environment and Climate Change Canada (the analog to the Environmental Protection Agency) projects a further loss of up to 60 percent by 2050—the result of declining snowfall and rising temperatures, both associated with a warming climate.

Of the years between 2003 and 2015, six qualified as drought years—that is, the Sierra snowpack came in woefully below the historic norm, leading to cuts in the allotment of water that farmers like Del Bosque received. During the 2011–17 drought, Del Bosque said, his water allotment went to zero for three years in a row, forcing him to buy expensive water from farmers in the wetter counties of the northern Central Valley—at prices that essentially wiped out his operation's profits.

University of California, Irvine, researchers found in a 2018 study that every 1-degree-Celsius rise in global average temperature leads to a 20 percent increase in the likelihood of a below-average Sierra Nevada snowpack. In its latest report, released in October 2018, the United Nations' Intergovernmental Panel on Climate Change projected at least a 0.5-degree-Celsius jump in temperature by as soon as 2030—foretelling a parched future for the San Joaquin Valley, and ever more challenges for the region as a pillar of our food supply.

When I visited him in 2018, a last-minute March blizzard in the Sierra Nevada partially saved what would have otherwise been another dismal snow year. But even with the late flurry, the mountain range's snowpack reached just 58 percent of its historic norm, and Del Bosque's farm initially received only 20 percent of its traditional water allotment—and he didn't get more until it was too late in the season to be used effectively. Again, he had to buy expensive water.

Ideally, farmers like Del Bosque would have a backup groundwater source beneath their feet for dry years, with the aid of high-powered wells. But in Del Bosque's part on the valley's west side, the aquifer is already fairly parched. What little water exists is laden with salt: as aquifers deplete, their water gets more and more concentrated with naturally occurring minerals, including salts. Most plants, including almonds, don't thrive in salty water, so groundwater for him is a limited resource; to be usable, it has to be diluted with surface water, which farmers must mix together in an irrigation pond. Even in places where the groundwater is not too salty, it is now a dwindling resource after years of pumping the San Joaquin's aquifers to offset the deficit created by the declining Sierra Nevada snowpack.

Put all this together—booming markets for almonds and organic melons, mounting labor and water woes—and Del Bosque has a thriving business built on shaky ground, vulnerable to wholly external factors like immigration enforcement and weather patterns.

Del Bosque's issues around water and labor encapsulate this current dilemma: melons are a relatively water-efficient crop that faces increasingly high labor costs; and almonds are an extremely labor-efficient crop that takes a huge gulp out of an increasingly scarce water supply. Farmers throughout the San Joaquin Valley and California's other high-output growing regions face the same conundrum, which also amounts to a massive challenge for the American eater.

Walk into any supermarket—from San Diego to Seattle, from Houston to Chicago, from Miami to Bangor—and the produce section brims with goods from California operations like Del Bosque's. Altogether, California farms churn out more than one third of the vegetables and two thirds of both fruits and nuts grown in the United States, the bulk of them in the Central Valley. Look at the numbers for specific commodities, and you get a jaw-dropping picture: a veritable roster of foods we're advised to eat more of. California produces nearly all of the almonds, walnuts, and pistachios consumed domestically; 90 percent or more of the broccoli, carrots, garlic, celery, grapes, tangerines, plums, and artichokes; at least 75 percent of the cauliflower, apricots, lemons, strawberries, and raspberries; and more than 40 percent of the lettuce, cabbage, oranges, peaches, and peppers.

And as if that weren't enough, the state is also a national hub for milk production. Tucked in amid the almond groves and vegetable fields of the San Joaquin Valley are vast dairy operations that confine cows together by the thousands and produce more than a fifth of the nation's milk supply, more than any other state. ("We got one about a mile east of here and another four miles south," Del Bosque told me. "Luckily, they're downwind.") California is also the nation's number one producer of the main feed for all those dairy cows: alfalfa, a water-intensive hay crop.

It all amounts to a food-production juggernaut: California generates $49 billion worth of food per year—loads of milk and beef in addition to fresh produce and nuts—nearly double the haul of its closest competitor among U.S. states, the corn-and-soybean behemoth Iowa. California's labor issues are a major concern, and intimately affect the lives of workers, owners, and consumers alike. But on that score, a change in political winds could make a big difference for farmers like Del Bosque. Harder to reverse, though, are troubling developments in the region's ecology, developments whose roots go back to the earliest days of human settlement in California, and whose effects today are running down the state's groundwater and surface water supplies faster than they can be replenished.

So where will the water come from to keep this juggernaut rolling?

Bordered on all sides by mountains, the Central Valley stretches 450 miles long, is on average 50 miles wide, and occupies a land mass of 18,000 square miles, or 11.5 million acres—roughly equivalent in size to Massachusetts and Vermont combined. Wedged between the Sierra Nevada to the east and the Coast Ranges to the west, it's one of the globe's greatest expanses of fertile soil and temperate weather. It's divided into two sections: The drier San Joaquin Valley, home to Del Bosque Farms, makes up the southern two thirds—from the southern border of the Tehachapi Mountains, south of Bakersfield, up to Sacramento. The slightly lusher Sacramento Valley makes up the northern third, extending from Sacramento to the Cascade Mountains.

Apart from budding metropolises like Fresno and Modesto—which have emerged as magnets for workers priced out of the stratospheric housing markets in Los Angeles and the San Francisco Bay Area—virtually the entire valley bottom is a flat, dusty expanse, carpeted with crops for most of the year. Drive through the San Joaquin Valley during the summer and you'll cross bridges over wide, dry gashes in the land: phantom rivers cut by millennia of snowmelt that are now diverted into irrigation canals most of the year. The vanished water is embedded in the crops you'll see flourishing all around you (and the produce that stocks your fridge at home).

The Central Valley's emergence as an agricultural epicenter utterly transformed the landscape. Before California became a U.S. state in 1850, rivers "flowed uninterrupted into valleys, marshes, bays, and the ocean," the late environmental historian Norris Hundley says in his masterful history *The Great Thirst: Californians and Water.* The indigenous people who thrived in the San Joaquin Valley "went to the water sources and settled near them, rather than diverting water or storing it," Hundley writes. And water sources were plentiful. The Central Valley was the site of "numerous rivers, lakes, and marshlands that were in existence more or less year-round and alternately expanded or contracted with the rhythm of the seasons." The waterways teemed with fish and supported a variety of land animals, including beavers and otters. Water-adapted plants, too, proliferated—cattails, tules (a ubiquitous marsh grass), willows, and alder—providing habitat for ducks, swans, marsh wrens, rails, and geese "that darkened the skies with their enormous numbers," Hundley writes:

> At least 4 million acres of the Central Valley, more than a third of its landmass, were natural wetlands, giving ample refuge to migratory birds along the Pacific flyway. Wetlands also served as the conduit between above-ground water flows and underground aquifers, mitigating floods and filtering water as it seeps downward.

The mighty Sacramento River, the largest in California, flows south from its headwaters at Mount Shasta, in the Cascade Mountains, the Central Valley's northern border. The San Joaquin River flows west and north from the Sierra Nevada. The two rivers ultimately converge just south of the state capital of Sacramento and create the Sacramento–San Joaquin River Delta, the largest estuary on the west coast, at 738,000 acres, covering more than 40 percent of the landmass in California. The stunningly biodiverse and productive ecosystem drains into the Pacific at the Golden Gate in San Francisco Bay. Before "man-made structures dictated otherwise," the rivers "constantly shaped and reshaped the contours of the great valley," Hundley writes. At the time of European contact, salmon was so abundant that Native American fishers within the basin harvested 8.5 million pounds or more of the nutrient-dense fish annually.

Farther south, the Kern, Kings, Tule, and Kaweah Rivers drained into a low-lying basin, never reaching the sea, generating enormous lakes. Tulare Lake, then the largest body of fresh water west of the Mississippi, teeming with fish and turtles, occupied a swath of land in the southern San Joaquin Valley four times the size of present-day Lake Tahoe. To the south of Tulare

Lake, two more now-vanished lakes, Buena Vista and Kern, would periodically "combine into a single large body of water, following an especially large snowmelt in the Sierra Nevada," Hundley reports.

These lake-river-wetland ecosystems, replenished annually by snowmelt, supported a spectacular diversity of plant and animal life, from oak trees and elk to mussels and tules. People, too, flourished. The nineteen Tulare Basin–dwelling tribes of the Yokut Nation had a combined population of at least nineteen thousand by the time of first contact with Spanish colonial invaders in 1772—the densest non-agricultural population in the pre-Columbian North American continent, according to California Polytechnic State University historical geographer William L. Preston.

While the Yokut didn't divert water flows to irrigate crops, they did devise "ways to make basin habitats more productive," Preston writes, accumulating food surpluses through "burning, plant care, and the establishment of boundaries, settlements, and pathways." Their diet was plentiful and varied: acorns from oak forests provided a protein- and fat-rich staple, supplemented by elk, small game, fish and shellfish, grass seeds, and as many as one hundred species of vegetables.

Meanwhile, the flows of the Central Valley's great rivers literally laid the groundwork for the modern-day agricultural boom: "Over the millennia, the rivers and their tributaries during the flood stage carried enormous quantities of topsoil from upland areas and deposited this rich silt across the floodplain that at the time of European contact constituted the nearly level valley floor," writes Hundley.

The region's pre-contact ecology set the stage for today's agriculture boom in another way, too: by stocking it with underground water. All those lakes, rivers, tributaries, and wetlands meant water percolating into aquifers everywhere except for the valley's arid western zones. "The ground, in effect, was filled to the brim," Hundley writes. Artesian springs proliferated, serving as "surface expressions of the vast storehouse of groundwater that helped nourish the abundant flora and fauna on the surface."

Today's almond, pistachio, and grape plantations are like a funhouse-mirror reflection of what Hundley called the "dense and sometimes impenetrable forests of willow, sycamore, oak, elder, poplar, alder, and wild grape" that thrived in the zones bordering rivers and lakes, providing "shelter and food for a rich array of wildlife."

The region's Spanish and later Mexican colonizing settlers didn't find much agricultural or urban use for the Central Valley. Ultimately giving up the idea of establishing inland missions, the Spanish saw it mainly as an unruly hiding place for indigenous people fleeing the crown's coastal settlements. They made violent forays into the valley to capture escapees, spreading disease, upending the social order, and triggering cultural unraveling. The genocide decimated the valley's original inhabitants. Between 1769, when the Spanish first claimed control of Alta California, and 1849, when U.S. settlers seized the territory, the indigenous population had plunged to about a quarter of its pre-contact size in the greater San Joaquin Valley. In the Tulare Basin, the number of native people had dwindled to some two thousand; for every ten original inhabitants, only about one remained.

The landscape, meanwhile, had changed, too: grasslands that once supported large elk populations had in places been overrun by cattle and horses owned by California's Mexican rancheros, some through intentional grazing and some after having escaped coastal ranches. Their presence essentially obliterated the native grass species that had helped build up the region's soil over millennia, and provided a home for native fauna, which declined as well.

But Spanish and Mexican rule didn't interfere with the free flow of water from the Sierra Nevada into and through the valley; dam- and canal-building were virtually nonexistent in the first two hundred years of colonial settlement. All that changed in 1849, with the discovery of gold in the Sierra Nevada foothills due east of present-day Sacramento. The short-lived gold rush drew a massive wave of white settlers from the United States, which had seized hold of the California territory after the Mexican-American War. California formally became a U.S. state in 1850. As the gold rush went bust, settlers began to focus on seizing Central Valley land, altering its waterways to raise cattle and grow wheat, cotton, and ultimately fruits and vegetables for the new state's fast-growing coastal population.

In their lunge for quick riches, they invented a principle water use knows as "prior-appropriation" rights. It meant that the first settler to divert a stream and put it to "beneficial use"—for example, for gold extraction—established a right to continue to extract that quantity of water in perpetuity. (Indigenous people were, of course, excluded from these rights.) As for groundwater, a settler who established ownership also had the right to any water that could be obtained from below. Astonishingly, prior appropriation, defined by the phrase "first in time, first in use," forms the basis for water law in California and much of the West to this day.

Meanwhile, the remaining Yokuts of Tulare Lake resisted Anglo settlement on their lands. Among the first outposts was Visalia, then a creek-crossed village at the eastern edge of the lake, now an agribusiness and farmworker hub in bone-dry Tulare County. Whereas the Spanish had viewed the native population as potential religious converts and tax-paying subjects of the crown, U.S. settlers saw them as an obstacle to be removed. They mobilized to expel the Yokut or, failing that, exterminate them. After seven years of violent hostilities, in 1857 the state government consigned the region's remaining indigenous people to live on a reservation in the present-day Tulare County town of Porterville. To create more distance between themselves and the indigenous people they had displaced, the settlers ultimately moved the reservation twenty miles east, naming it the Tule River Tribe Reservation in the Sierra foothills. Today, the reservation has a population of 1,600, with an additional 3,000 tribal members living nearby.

The San Joaquin Valley landscape of today would be unrecognizable to a Yokut or white colonialist circa 1850. A $49 billion desert-agriculture empire has arisen on what was once a vibrant, river-crossed homeplace, or a vast soggy void in desperate need of improvement, depending on the observer's point of view. A band of wetlands that once covered about 30 percent of the Central Valley floor has essentially vanished, its water sources diverted away. Today, just about 380,000 acres of wetland remain, about 9 percent of the nearly 4 million acres that existed in the 1850s—and that was preserved only after a heroic push for restoration by environmentalists horrified by the destruction of crucial habitat for wild birds along the Pacific Flyway.

The Kern River and its tributaries, which once conveyed Sierra Nevada snowmelt to the web of lakes and wetlands that characterized the Tulare Basin, now flow into dams and canals, bringing water to farms on the parched basin floor. Farther north, as much as 80 percent of the flow of the San Joaquin River, which fed its own swath of wetlands as it meandered north to the Sacramento–San Joaquin River Delta, is now diverted into irrigation networks; so is nearly 40 percent of the Sacramento River, which runs through the northern third of the Central Valley.

By the early 1970s, after a century of ad hoc efforts by valley farmers, and a government dam-building boom that began in the 1930s, these waterways were finally fully developed and manipulated. Credit for this transformation ultimately goes to a great mobilization of political will by farmers. Two multibillion-dollar government initiatives—the federally funded Central Valley Project and the California-funded State Water Project—arose to capture and distribute the annual Sierra Nevada snowmelt, protecting farmland from floods while supplying it with on-demand water. Eighty percent of the state's "developed" water—the stuff that moves through human-made infrastructure—goes to agriculture. Combined, all the urban and suburban areas in the state use only 20 percent of California's developed water.

In years when Sierra Nevada snow is abundant, the Central Valley is a near-ideal place to grow the crops that stock the produce aisle and the bulk bins at the grocery store. The region's precious little rainfall occurs almost exclusively during the mild winters; in the long, hot growing season, farmers can essentially control irrigation to the last drop, giving their crops what they need without the diseases and pest pressure that come with summer rains. Even in low-snow years, most farmers can revert to water from underground to water their crops.

From the era of California's statehood to the present, water mastery has played a pivotal role in constructing the Central Valley's food-production machine. Access to a cheap migrant labor force has always been a factor—and remains so today for growers like Del Bosque. But nothing is possible without water, and the post–gold rush land grabs were essentially water grabs, too. The valley's position below the Sierra Nevada made it a sponge for a portion of every year's snowmelt, giving it enormous reserves of aquifer water. Under the property law governing California from statehood until 2040—when legislation passed in 2014 finally goes into effect—landowners are free to use the water resources beneath their feet as they wish, and to take from passing waterways according to privileges that were packaged with land deeds. Exploitation of water drove the valley's rise as the nation's produce patch.

By the 1920s, with the onset of electrification, a veritable well-drilling frenzy among San Joaquin Valley farmers triggered an aquifer extraction so vast that the land literally began to sink. The scientific term for this phenomenon is "subsidence"; it's what happens as water is pumped out from deep wells, causing soil to settle in uneven and unpredictable ways. In a 1975 report, the U.S. Geological Survey (USGS) observed that subsidence in the San Joaquin Valley "represents one of the great changes man has imposed on the environment." The authors continue:

> About 5,200 square miles of irrigable land, one-half the entire valley, has been affected by subsidence, and maximum subsidence exceeded 28 feet in 1970; by 1972 subsidence was about 29 feet. Throughout most of the area, subsidence has occurred so slowly and over such a broad area that its effects have gone largely unnoticed by most residents.

Although sinking at rates as high as ten inches per year may be hard to perceive as it happens, its effects are ultimately impossible to ignore.

Subsidence is the enemy of built infrastructure; it causes bridges and irrigation canals to buckle and malfunction; it ruptures the casings of wells. More important, as surface land sinks, aquifers shrink in size, meaning they have less capacity to store water in the future. Every inch of subsidence means less potential for future water storage.

But the USGS report brought good news, too. As of 1973, the report stated, "after three decades of continued declining water levels, many hundreds of irrigation wells are idle and water levels are rising." As a result, "throughout much of the valley, artesian pressures are recovering toward their pre-subsidence levels, and elevations of the subsiding land surface are stabilizing." The reason for the turnaround: those state and federal projects conveying Sierra Nevada snowmelt were finally in place, reliable, and fully operational, providing an annual dose of irrigation water that kept the otherwise arid bottomlands in bloom without the need to pump aquifers dry. California had created what the eminent historical geographer Richard Walker called the "world's largest water storage and transfer system, consisting of some 1,600 major dams and thousands of miles of canals and aqueducts." The valley's water needs had been brought into balance—and the bounty it delivered the nation was secure.

That 1975 USGS report implied a kind of aquatic utopia: a vision for a plentiful and stable U.S. food supply. Here's how it was supposed to work. Most years, the big irrigation projects would distribute annual doses of snowmelt that would more than meet valley farmers' needs. During inevitable drier-than-normal years, farmers would receive less water from the projects and make up for it by tapping their aquifers. In financial terms, the snowmelt is supposed to function like a regular irrigation paycheck; the aquifers, as a kind of underground savings account. When winter snows returned in full force, farmers would lay off their wells, and the projects would deliver enough snowmelt to both irrigate their fields and recharge their aquifers, stocking them up for the next dry year.

But the twentieth century—when California's food-production colossus arose, when the public purse, guided by the influence of agribusiness, opened to build out the world's greatest water-conveyance system—turns out to have been an unusually wet and stable one, weather-wise.

Not that it was without epic droughts. The Dust Bowl period of the 1930s is most known for the brutal dry spell that parched the Southern Plains a thousand miles east of California; but the Sierra Nevada, too, received tiny amounts of snow in the years between 1927 and 1935, curtailing the amount of groundwater flowing through the valley and helping spark the first well-drilling bonanza.

In 1976—just a year after that sanguine USGS report—a severe two-year drought set in, forcing the big government water projects to slash deliveries to San Joaquin Valley farms and triggering another big drawdown of the region's aquifers. The water table plunged by fifty feet, and farmers drilled or deepened nine thousand wells in just two years. A decade later, another multiyear drought set in, lasting from 1987 to 1992, triggering yet another major water extraction from beneath the valley.

This brings us back to Joe Del Bosque and his water dilemma. So much water has been extracted over the course of the new century that the ground has begun sinking again—in some places by as much as two feet per year.

Between 1998 and 2014, a period that included two major droughts, wells dug into San Joaquin Valley aquifers produced an average of 3.5 cubic kilometers of water per year, according to a 2015 USGS report. (By comparison, the entire city of Los Angeles uses about 0.67 cubic kilometers annually.) In other words, San Joaquin Valley farmers have become so ravenous for water that they are using far more than is annually generated by the Sierra Nevada snowpack, and they must revert to the well pump to make up the difference.

If relying on annual snowmelt is like living off your paycheck, relying on groundwater is akin to prematurely raiding your 401(k). Every draft you take is one that you won't be able to replenish, at least not easily or cheaply.

As you'd expect, underground water storage drops during dry years, as farmers must use the pump to make up for lost irrigation allotments; and it rises during wet years, when the irrigation props up their contribution. The problem is, aquifer recharge during wet years never fully replaces all that was taken away during dry times.

Meanwhile, the titanic transfer of water from underground and into crops and ultimately the food supply has left quite a void in its wake. By the end of the 2011–17 drought, large swaths of the San Joaquin Valley were sinking by as much as two feet per year. One subsidence hot spot identified by the NASA Earth Observatory, using satellite data from May 2015 to September 2016, is the town of Tranquility, California, twenty miles southeast of Firebaugh, where Del Bosque farms. Another, the epicenter of the subsidence bowl, is Corcoran (eighty miles southeast of Tranquility), which just 150 years ago would have been submerged at the bottom of Tulare Lake.

In addition to damaging roads, bridges, houses, and pretty much any built infrastructure, subsidence snarls up the canals that carry snowmelt from the Sierra Nevada. These state and federally maintained projects are a feat of engineering, taking advantage of the huge altitude drop from the mountain range to carry water via gravity to farms across the valley. Indentations of just a few inches are enough to impede flow; the multiple-foot drops of recent years are a disaster. The California Aqueduct is the main artery of the State Water Project—it carries surface water to a million acres of San Joaquin Valley farmland. The California Department of Water Resources disclosed in 2017 that recent subsidence, in some places as deep as twenty-five inches, had reduced the canal's maximum flow (read: increased water lost to leaks) by 20 percent. A similar impediment afflicts the Delta-Mendota Canal, a major artery of the Central Valley Project, a federally run network of dams, reservoirs, and canals that water about a third of California's irrigated farmland and provide water and electricity to millions of urban users, all from snowmelt; the Friant-Kern Canal, another crucial channel that brings water to the Tulare Basin, has seen its flow reduced by 60 percent because of subsidence-related damage.

Thus, we have a vicious circle: reduced snowmelt means less water flowing through government-run irrigation channels, which pushes farmers to pump more water from underground, causing subsidence that damages those channels and reduces their flow capacity, pushing farmers to accelerate the cycle by pumping more water from underground.

No one knows exactly how much water we have left. When I interviewed Jay Famiglietti, a senior water scientist at NASA and executive director of the University of Saskatchewan's Global Institute for Water Security, he pointed out that very little research is done on how much underground water there is, unlike the amount of research on underground oil. Given how much the U.S. food system relies on California, "we sure as heck need to quantify the available volume of freshwater," he said.

Recent events could seem tame compared with what's to come. The epochal drought of 2011 to 2017, the worst in California's recorded history—falling fast upon an only slightly less severe dry spell that lasted from 2007 to 2009—could prove to be close to a new norm.

Erratic as it is, California's weather follows a pattern, one it shares with other Mediterranean-climate regions, which are clustered at the western edge of continents at middle latitudes. The basic template: dry, warm summers and wet, cool winters. (In California, summers are milder at the coast, tempered by the Pacific, and blazing-hot in the Central Valley.)

The driver for the wet season–dry season dynamic is what climatologists call the "semipermanent high-pressure systems" that settle between subtropical regions and the poles. These high-pressure zones move poleward during the summer, meaning that storms during that season are pushed northward (most famously over sodden Seattle and the rest of the Pacific Northwest), leaving California mostly dry. Over winter, the high-pressure concentration recedes, lurching southward, opening the California coast to storms from the western Pacific. As these storms move east over land, they hit the Sierra Nevada, where they're forced upward into the cold temperatures of the high altitudes and shed their moisture in the form of snow.

So California has a relatively tight window to receive the great bulk of its precipitation: from November through March. This pattern helps explain the state's vulnerability to droughts. The difference between a bountiful water year and a dry one is thin; the failure of a large storm or two to arrive before the seasonal window slams shut can make all the difference.

In December 2012, the high-pressure zone didn't follow its normal pattern of receding over winter. Instead, it lingered, pushing the Pacific jet stream to the north of California, shielding the California coast and leading to a near-complete failure of a Sierra Nevada snowpack to develop. A year later, when the pressure zone again showed no signs of moving along its normal path, Donald Swain, a climate scientist at the University of California, Los Angeles, dubbed it the "ridiculously resilient ridge" (the so-called Triple R), with "ridge" being meteorologist-speak for a pocket of high atmospheric pressure.

Swain wrote that at the time, he assumed it would dissipate within a few weeks. Instead, "the 'Triple R' held strong straight through the entire winter—and then recurred, in slightly modified form, throughout the winters of 2014–2015 and 2015–2016." He continued:

> The multi-year persistence of this anomalous atmospheric ridge was nothing short of extraordinary. The co-occurrence of record low precipitation and record high temperatures associated with the Triple R ultimately yielded California's most severe multi-year drought on record.

Inspired to look at recent decades' meteorological data for evidence of similar weather patterns in the past, Swain and a team of researchers from Stanford and Columbia Universities found a disturbing context for the epic drought. While the 2011–17 situation was indeed unprecedented, they found, previous dry spells over the preceding half century had also shown pressure ridges that persisted into winter; the more extreme the ridge, the bigger the drought. And such events showed a clearly increasing trend over the decades, due in no small part to factors related to climate change: warmer temperatures within and above the Pacific Ocean as well as rising sea levels.

In short, the ridiculously resilient ridge that settled along the California coast likely won't seem quite so remarkable as the climate continues warming. And in a classic feedback loop, warmer temperatures will mean that severe droughts will get even worse. In a 2018 paper, researchers from Lawrence Berkeley National Laboratory and the USGS modeled what such droughts will be like in a warmer mid-twenty-first century climate. Their finding: "increased soil drying," "many more extreme heat days" in the Central Valley (i.e., days above 104 degrees), and "record-low snowpack and record-high forest mortality" in the Sierra Nevada.

In another project designed to paint a picture of Central Valley agriculture in a warming future, the same group of researchers looked at the headwaters feeding the ten major reservoirs designed to capture snowmelt from the Sierra each year, using averages from 1985 to 2005 as a baseline.

Assuming global greenhouse gas emissions continue rising at present rates—that is, a "business as usual" scenario with no effective global deal to cut greenhouse gas emissions and no major technological breakthroughs—they applied nine different climate scenarios to model future flow.

The results: By mid-century (2039–59), the average annual snowpack will fall by 54.4 percent, compared with the late-twentieth century baseline. By the final decades of the century, when today's teens are in their seventies, the snowpack will be 79.3 percent beneath the standard enjoyed when California's farm infrastructure was built. To analyze massive amounts of water, planners think in acre-feet—the amount needed to submerge an acre of land by one foot. At the end of the last century, the Sierra Nevada captured an average of 8.76 million acre-feet. By mid-century, they project, the average will fall to 4 million acre-feet; and by century's end, 1.81 million acre-feet.

The Central Valley Project could become what economists called a "stranded asset" in such a scenario: a multibillion-dollar public investment that lacks sufficient snowmelt to perform its tasks.

As the authors note, these findings dovetail with those of California's Fourth Climate Change Assessment, released in August 2018: "As a result of projected warming, Sierra Nevada snowpacks will very likely be eradicated below about 6,000 feet elevation and will be much reduced by more than 60% across nearly all of the range."

Looking to the deep past provides no more comfort. California's fossil record depicts a climate that whiplashes between extreme wet and dry periods. In their 2013 book *The West Without Water: What Past Floods, Droughts, and Other Climate Clues Tell Us About Tomorrow*, University of

California, Berkeley, paleoclimatologist B. Lynn Ingram and environmental planner Frances Malamud-Roam provide a chilling summary of ancient drought science. They synthesize research on sediment cores at the Sacramento–San Joaquin River Delta, tree rings in foothills forests, and ancient tree stumps that appear in the lakes at the base of the Sierra Nevada during droughts to tell a story of drier times in the past.

Between the years 900 and 1400—a 500-year period ending 150 years before Europeans set foot in present-day California—the region experienced a series of decades-long "megadroughts," interrupted by a century-long wet interval starting around 1100. The archaeological record shows that indigenous peoples in the region reacted flexibly to the upheavals, moving to the coast during long dry stretches, flourishing in the valley during lush periods, and heading to the highlands during floods.

Weather patterns over the past twenty years suggest the region could be entering another long-term dry phase, Ingram told me in an interview. "The twentieth century was a relatively wet time—and a time when all of our modern societies were built, including California's agricultural infrastructure," Ingram said. "We've had centuries where it was far drier. We're not prepared."

Layering anthropomorphic climate change onto this deep history does not provide comfort. According to a paper by researchers from the NASA Goddard Institute for Space Studies, Columbia University, and Cornell University, California in the second half of this century will "likely exceed even the most severe megadrought periods of the Medieval era." The prediction holds even if global greenhouse gas emissions see a modest decline in the next two decades, the study's authors write: "Our results point to a remarkably drier future that falls far outside the contemporary experience of natural and human systems in Western North America, conditions that may present a substantial challenge to adaptation."

Ingram's work demonstrates that the California territory is dominated by cycles of extreme weather that predate European contact by centuries. The region's phase as a U.S. state has been too short for most residents to experience the worst shocks that characterize its particular climate. An alarming accumulation of research suggests that climate change will only amplify the chaos.

The bad news is piling up at a time when Central Valley farmers (egged on by the Trump administration) are already fighting with state managers for larger annual allocations of snowmelt, at least some of which has to flow into the Sacramento–San Joaquin River Delta and to the Pacific. Campaigning for the presidency in the summer of 2016, at the very height of the long drought, Trump assured a cheering crowd in Fresno that "there is no drought." The real factor parching their farms, he insisted, was the jackboot of government regulation, pressed into the service of mindless environmentalism. "You have a water problem that is so insane!" he thundered. "It is so ridiculous, where they're taking the water and shoving it out to sea." The candidate vowed to "solve" their water problem through deregulation. The crowd erupted in cheers.

On cue, in fall 2019, Trump's Department of the Interior made a decision that farming interests in the San Joaquin Valley had long clamored for. The agency effectively gutted Endangered Species Act protections for Delta-dwelling fish species, a change that will allow large amounts of water to be diverted to irrigate San Joaquin Valley farmland. Interior Secretary David Bernhardt, who championed the decision, had previously worked as a lobbyist for the Westlands Water District, a state-chartered entity that represents large-scale farms on the San Joaquin Valley's parched west side.

While the increased water diversions won't be nearly enough to solve west side agriculture's water deficit, they could well spell doom for a vast and important ecosystem. According to the Water Education Foundation, "Eighty percent of the state's commercial fishery species live in or migrate through the Delta, and at least half of its Pacific Flyway migratory water birds rely on the region's wetlands." Even before Trump's move, its ecosystem already teeters near collapse, driven in part by the relentless shunting of water to farm interests in the Central Valley.

Meanwhile, the more water that gets pumped from underground aquifers, the lower that water tables drop, meaning that wells have to go deeper into the earth, increasing pumping costs. Water from deep underground tends to be higher in minerals, including those salts that settle in the soil and hinder the ability of plants to take up water, decreasing productivity. Already, the San Joaquin Valley's soils are accumulating salts at an alarming rate, costing California farmers $3.7 billion in lost crop yields, according to one 2017 study. The great bulk of the damage is taking place along the arid western side of the valley, farther away from the Sierra Nevada snowmelt.

Take a trip through the rural roads of the area and you'll see it: groves with wan, suffering almond trees rising up from soil caked with visible white salt sparkling in the sun. Del Bosque told me that the ground he farms is getting salty, too—just one more thing to worry about.

The legal requirement to recognize that farmworkers must have the same rights as all other workers, the vanishing water, the creeping salinization—it's a lot for Del Bosque to contend with. "So, yeah, I'm hoping to pass the farm on to my kids," he told me toward the end of our conversation on a hot September day. "But there are some uncertainties here."

2

The Flood Next Time

In November 1860, a young scientist from upstate New York named William Brewer disembarked in San Francisco after a long journey that took him from New York City through Panama and then north along the Pacific coast. "The weather is perfectly heavenly," he enthused in a letter to his brother back east. "They say this is a fair specimen of winter here, yet the weather is very like the finest of our Indian summer, only not so smoky—warm, balmy, not hot, clear, bracing." The fast-growing metropolis was already revealing its charms: "large streets, magnificent buildings of brick, and many even of granite, built in a substantial manner, give a look of much greater age than the city has." He described San Francisco as a kind of gorgeous urban garden, graced by "many flowers we [northeasterners] see only in house cultivations: various kinds of geraniums growing of immense size, dew plant growing like a weed, acacia, fuchsia, etc. growing in the open air."

Brewer was on a serious mission. Barely a decade after being claimed as a U.S. state, California was plunged into an economic crisis. The gold rush had gone bust, and thousands of restive U.S. settlers were left scurrying about, hot after the next ever-elusive mineral bonanza. The fledgling legislature had seen fit to hire a state geographer to gauge the mineral wealth underneath its vast and varied terrain, hoping to organize and rationalize the mad lunge for buried treasure. The potential for boosting agriculture as a hedge against mining wasn't lost on the state's leaders. They called on the state geographer to deliver a "full and scientific description of the state's rocks, fossils, soils, and minerals, and its botanical and zoological productions, together with specimens of same." The task of completing the fieldwork fell to the thirty-two-year-old Brewer, a Yale-trained botanist who had studied cutting-edge agricultural science in Europe. His letters home, chronicling his four-year journey up and down California, form one of the most vivid contemporary accounts of its early statehood.

They also provide a stark look at the greatest natural disaster known to have befallen the western United States at least since European contact in the sixteenth century.

In November 1861, barely a year after Brewer's sunny initial descent from a ship in San Francisco Bay, he was back in the city, on break after a year of surveying. He complained in a letter home of a "week of rain." In his next letter, two months later, Brewer reported jaw-dropping news: rain had fallen almost continuously since he had last written—and now the entire Central Valley was underwater. "Although much of it is not cultivated, yet a part of it is the garden of the state," he wrote. "Thousands of farms are entirely underwater—cattle starving and drowning."

Picking up the letter nine days later, he wrote that a bad situation had deteriorated: "We have had very bad weather since the above was written." All the roads in the middle of the state are "impassable, so all mails are cut off." Telegraph service, which had only recently been connected to the East Coast through the Central Valley, stalled. "*The tops of the poles are under water!*" The young state's capital city, Sacramento, about a hundred miles northeast of San Francisco at the western edge of the valley and the intersection of two rivers, was submerged, forcing the legislature to evacuate—and delaying a payment Brewer needed to forge ahead with his expedition.

The surveyor gaped at the sheer volume of rain. In a normal year, Brewer reported, San Francisco received about twenty inches. In the ten weeks leading up to January 18, 1862, the city got "thirty-two and three-quarters inches and it is still raining!" Meanwhile, "generally twice, sometimes three times, as much [has fallen] in the mining districts on the slopes of the Sierra"—some places getting as much as "*six feet*" in that span. He offered a stunned comparison to his rain-soaked homeplace, upstate New York: "As much rain as falls in Ithaca in two years has fallen in some places in this state in two months."

Brewer went on to recount scenes from the Central Valley that would fit in a Hollywood disaster epic. "An old acquaintance, a *buccaro* [cowboy], came down from a ranch that was overflowed," he wrote. "The floor of their one-story house was six weeks under water before the house went to pieces." Steam boats "ran back over the ranches fourteen miles from the [Sacramento] river, carrying stock [cattle], etc., to the hills," he reported. He marveled at the massive impromptu lake made up of "water ice cold and muddy," in which "winds made high waves which beat the farm homes in pieces." As a result, "every house and farm over this immense region is gone."

Eventually, in March, Brewer made it to Sacramento, hoping (without success) to lay hands on the state funds he needed to continue his survey. He found a city still in ruins, weeks after the worst of the rains. "Such a desolate scene I hope never to see again," he wrote:

> Most of the city is still under water, and has been for three months . . . Every low place is full—cellars and yards are full, houses and walls wet, everything uncomfortable. Over much of the city boats are still the only means of getting about . . . Houses, stores, stables, everything, were surrounded by water. Yards were ponds enclosed by dilapidated, muddy, slimy fences; household furniture, chairs, tables, sofas, the fragments of houses, were floating in the muddy waters or lodged in nooks and corners—I saw three sofas floating in different yards.

The "better class of houses" were in rough shape, Brewer observed, but "it is with the poorer classes that this is the worst." He went on: "Many of the one-story houses are entirely uninhabitable; others, where the floors are above the water are, at best, most wretched places in which to live." He summarized the scene:

> Many houses have partially toppled over; some have been carried from their foundations, several streets (now avenues of water) are blocked up with houses that have floated in them, dead animals lie about here and there—a dreadful picture. I don't think the city will ever rise from the shock, I don't see how it can.

Water scarcity, it turns out, isn't the only menace that stalks the California valleys that stock our supermarkets. The opposite—catastrophic flooding—also occupies a niche in what Mike Davis, the great chronicler of Southern California's sociopolitical geography, has called the state's "ecology of fear." Indeed, his classic book of that title opens with an account of a 1995 deluge that saw "million-dollar homes tobogganed off their hill-slope perches" and small children and pets "sucked into the deadly vortices of the flood channels."

Yet floods tend to be less feared than rival horsemen of the apocalypse in the state's oft-stimulated imagination of disaster. The epochal 2011–17 drought, with its missing-in-action snowpacks and draconian (for some) water restrictions, burned itself into the state's consciousness. Californians are rightly terrified of fires like the ones that roared through the northern Sierra Nevada foothills and coastal canyons near Los Angeles in the fall of 2018, killing nearly one hundred people and fouling air for miles around; many are frightfully aware that a warming climate will make such conflagrations increasingly frequent. And "earthquake kits" are common gear in closets and garages all along the San Andreas Fault, where the next "Big One" lurks. Floods, though they occur as often in Southern and Central California as they do anywhere in the United States, don't generate quite the same buzz.

But a growing body of research shows there is a flipside to the megadroughts Central Valley farmers like Del Bosque face: megafloods. In a region that can sometimes seem like a "Book of the Apocalypse theme park" (Davis again), the nearly forgotten biblical-scale flood documented by Brewer's letters has largely vanished from the public imagination. The event, known as the Great Flood of 1862, was once thought of (when it was thought of at all) as a thousand-year anomaly, a freak occurrence. But emerging science demonstrates that floods of much greater magnitude occurred every one hundred to two hundred years in California's precolonial history—and climate change will likely make them more frequent still. The region most vulnerable to such a water-drenched cataclysm in the near future is, ironically enough, the San Joaquin Valley's arid, sinking basin, the beleaguered jewel of the U.S. food system.

At the time of the Great Flood, the Central Valley was still mainly cattle ranches, the farming boom a ways off. Late in 1861, the state suddenly emerged from a two-decade dry spell when monster storms began lashing the west coast from Baja California to present-day Washington State. In central California, the deluge initially took the form of ten to fifteen feet of snow dumped onto the Sierra Nevada, according to research by the U.C. Berkeley paleoclimatologist B. Lynn Ingram, who has emerged as a kind of Cassandra of flood preparedness in the western United States. Soon after the blizzards came days of warm, heavy rain, which in turn melted the enormous snowpack. The resulting slurry cascaded through the Central Valley's network of untamed rivers.

As floodwater gathered in the valley, it formed a vast, muddy, wind-roiled lake, its size "rivaling that of Lake Superior," covering the entire Central Valley floor, from the southern slopes of the Cascade Mountains near the Oregon border to the Tehachapis, south of Bakersfield, with depths in some places exceeding fifteen feet.

At least some of the region's remnant indigenous population saw the epic flood coming and took precautions to escape devastation, Ingram reports, quoting an item in the *Nevada City Democrat* on January 11, 1862:

> We are informed that the Indians living in the vicinity of Marysville left their abodes a week or more ago for the foothills predicting an unprecedented overflow. They told the whites that the water

> would be higher than it has been for thirty years, and pointed high up on the trees and houses where it would come. The valley Indians have traditions that the water occasionally rises 15 or 20 feet higher than it has been at any time since the country was settled by whites, and as they live in the open air and watch closely all the weather indications, it is not improbable that they may have better means than the whites of anticipating a great storm.

All in all, thousands of people died, "one-third of the state's property was destroyed, and one home in eight was destroyed completely or carried away by the floodwaters," Ingram reports.

As for farming, the 1862 megaflood transformed valley agriculture, playing a decisive role in creating today's Anglo-dominated, crop-oriented Central Valley: a nineteenth-century example of the "disaster capitalism" that Naomi Klein describes in her 2007 book, *The Shock Doctrine.*

Prior to the event, valley land was still largely owned by Mexican rancheros who held titles dating to Spanish rule. The 1848 Treaty of Guadalupe Hidalgo, which triggered California's transfer from Mexican to U.S. control, gave rancheros U.S. citizenship and obligated the new government to honor their land titles. The treaty terms met with vigorous resentment from white settlers eager to shift from gold mining to growing food for the new state's burgeoning cities. The rancheros thrived during the gold rush, finding a booming market for beef in mining towns. By 1856, their fortunes had shifted. A severe drought that year cut production, competition from emerging U.S. settler ranchers meant lower prices, and punishing property taxes—imposed by land-poor settler politicians—caused a further squeeze. "As a result, rancheros began to lose their herds, their land, and their homes," writes the historian Lawrence James Jelinek.

The devastation of the 1862 flood, its effects magnified by a brutal drought that started immediately afterward and lasted through 1864, "delivered the final blow," Jelinek writes. Between 1860 and 1870, California's cattle herd, concentrated in the valley, plunged from 3 million to 630,000. The rancheros were forced to sell their land to white settlers at pennies per acre, and by 1870, "many rancheros had become day laborers in the towns," Jelinek reports. The valley's emerging class of settler farmers quickly turned to wheat and horticultural production, and set about harnessing and exploiting the region's water resources, both those gushing forth from the Sierra Madre and those beneath their feet.

Despite all the trauma it generated and the agricultural transformation it cemented in the Central Valley, the flood quickly faded from memory in California and the broader United States. To his shocked assessment of a still-flooded and supine Sacramento months after the storm, Brewer added a prophetic coda:

> No people can so stand calamity as this people. They are used to it. Everyone is familiar with the history of fortunes quickly made and as quickly lost. It seems here more than elsewhere the natural order of things. I might say, indeed, that the recklessness of the state blunts the keener feelings and takes the edge from this calamity.

Indeed, California's U.S. settlers ended up shaking off the cataclysm. What lesson does the Great Flood of 1862 hold for today? The question is important. Back then, just around 500,000 people lived in the entire state, and the Central Valley was a sparsely populated badland. Today, the valley has a population of 6.5 million people and boasts the state's three fastest-growing counties. Sacramento (population 501,344), Fresno (538,330), and Bakersfield (386,839) are all budding metropolises. The state's long-awaited high-speed train, if it's ever completed, will place Fresno residents within an hour of Silicon Valley, driving up its appeal as a bedroom community.

In addition to the potentially vast human toll, there's also the fact that the Central Valley has emerged as a major linchpin of the U.S. and global food system. Could it really be submerged under fifteen feet of water again—and what would that mean?

In less than two centuries as a U.S. state, California has maintained its reputation as a sunny paradise while also enduring the nation's most erratic climate: the occasional massive winter storm roaring in from the Pacific; years-long droughts. But recent investigations into the fossil record show that these past years have been relatively stable.

One avenue of this research is the study of the regular megadroughts, the most recent of which occurred just a century before Europeans made landfall on the North American west coast. As we are now learning, those decades-long arid stretches were just as regularly interrupted by enormous storms—many even grander than the one that began in December 1861. (Indeed, that event itself was directly preceded and followed by serious droughts.) In other words, the same patterns that make California vulnerable to droughts also make it ripe for floods.

Beginning in the 1980s, scientists including B. Lynn Ingram began examining streams and banks in the enormous delta network that together serve as the bathtub drain through which most Central Valley runoff has flowed for millennia, reaching the ocean at the San Francisco Bay. (Now-vanished Tulare Lake gathered runoff in the southern part of the valley.) They took deep-core samples from river bottoms, because big storms that overflow the delta's banks transfer loads of soil and silt from the Sierra Nevada and deposit a portion of it in the Delta. They also looked at fluctuations in old plant material buried in the sediment layers. Plant species that thrive in freshwater suggest wet periods, as heavy runoff from the mountains crowds out seawater. Salt-tolerant species denote dry spells, as sparse mountain runoff allows seawater to work into the delta.

What they found was stunning. The Great Flood of 1862 was no one-off black-swan event. Summarizing the science, Ingram and USGS researcher Michael Dettinger deliver the dire news: a flood comparable to—and sometimes much more intense than—the 1861–62 catastrophe occurred in each of the following time spans: 1235–1360, 1395–1410, 1555–1615, 1750–70, and 1810–20: "That is, one megaflood every 100 to 200 years." They also discovered that the 1862 flood didn't appear in the sediment record in some sites that showed evidence of multiple massive events—suggesting that it was actually smaller than many of the floods that have inundated California over the centuries.

During its time as a U.S. food-production powerhouse, California has been known for its periodic droughts and storms. But Ingram and Dettinger's work pulls the lens back to view the broader timescale, revealing the region's swings between megadroughts and megastorms—ones more than severe enough to challenge concentrated food production, much less dense population centers.

The dynamics of these storms themselves explain why the state is also prone to such swings. Meteorologists have known for decades that those tempests that descend upon California over the winter—and from which the state receives the great bulk of its annual precipitation—carry moisture from the South Pacific. In the late 1990s, scientists discovered that these "pineapple expresses," as TV weather presenters call them, are a subset of a global weather phenomenon: long, wind-driven plumes of vapor about a mile above the sea that carry moisture from warm areas near the equator on a northeasterly path to colder, drier regions toward the poles. And they carry so much moisture—often more than twenty-five times the flow of the Mississippi River, over thousands of miles—that they've been dubbed "atmospheric rivers."

In a pioneering 1998 paper, researchers Yong Zhu and Reginald E. Newell found that nearly all the vapor transport between the subtropics (regions just south or north of the equator, depending on the hemisphere) toward the poles occurred in just five or six narrow bands. And California, it turns out, is the prime spot in the western side of the northern hemisphere for catching them at full force during the winter months.

As Ingram and Dettinger note, atmospheric rivers are the primary vector for California's floods. That includes pre-Columbian cataclysms as well as the Great Flood of 1862, all the way to the various smaller ones that regularly run through the state. Between 1950 and 2010, Ingram and Dettinger write, atmospheric rivers "caused more than 80 percent of flooding in California rivers and 81 percent of the 128 most well-documented levee breaks in California's Central Valley."

Paradoxically, they are at least as much a lifeblood as a curse. Between eight and eleven atmospheric rivers hit California every year, the great majority of them doing no major damage, and they deliver between 30 and 50 percent of the state's rain and snow. But the big ones are damaging indeed. Other researchers are reaching similar conclusions. In a study released in December 2019, a team from the U.S. Army Corps of Engineers and the Scripps Institution of Oceanography found that atmospheric-river storms accounted for 84 percent of insured flood damages in the western United States between 1978 and 2017; the thirteen biggest storms wrought more than half the damage.

So the state—and a substantial portion of our food system—exists on a razor's edge between droughts and floods, its annual water resources decided by massive, increasingly fickle transfers of moisture from the South Pacific. As Dettinger puts it, the "largest storms in California's precipitation regime not only typically end the state's frequent droughts, but their fluctuations also cause those droughts in the first place."

We know that before human civilization began spewing millions of tons of greenhouse gases into the atmosphere annually, California was due "one megaflood every 100 to 200 years"—and the last one hit more than a century and a half ago. What happens to this outlook when you heat up the atmosphere by one degree Celsius—and are on track to hit *at least* another half-degree Celsius increase by mid-century?

That was the question posed by Donald Swain and a team of researchers at the UCLA's Department of Atmospheric and Oceanic Sciences in a series of studies, the first of which was published in 2018. They took California's long pattern of droughts and floods and mapped it onto the climate models based on data specific to the region, looking out to century's end.

What they found isn't comforting. As the tropical Pacific Ocean and the atmosphere just above it warm, more seawater evaporates, feeding ever bigger atmospheric rivers gushing toward the California coast. As a result, the potential for storms on the scale of the ones that triggered the Great Flood has increased "more than threefold," they found. So an event expected to happen on average every two hundred years will now happen every sixty-five or so. It is "more likely than not we will see one by 2060," and it could plausibly happen again before century's end, they concluded.

As the risk of a catastrophic event increases, so will the frequency of what they call "precipitation whiplash": extremely wet seasons interrupted by extremely dry ones and vice versa. The winter of 2016–17 provides a template. That year, a series of atmospheric-river storms filled reservoirs and at one point threatened a major flood in the northern Central Valley, abruptly ending the worst multiyear drought in the state's recorded history.

Swings on that magnitude normally occur a handful of times each century, but in the model by Swain's team, "it goes from something that happens maybe once in a generation to something that happens two or three times," he told me in an interview. "Setting aside a repeat of 1862, these less intense events could still seriously test the limits of our water infrastructure." Like other efforts to map climate change onto California's weather, this one found that drought years characterized by low winter precipitation would likely increase—in this case, by a factor of as much as two, compared with mid-twentieth-century patterns. But extreme-wet winter seasons, accumulating at least as much precipitation as 2016–17, will grow even more: they could be three times as common as they were before the atmosphere began its current warming trend.

While lots of very wet years—at least the ones that don't reach 1861–62 levels—might sound encouraging for food production in the Central Valley, there's a catch, Swain said. His study looked purely at precipitation, independent of whether it fell as rain or snow. A growing body of research suggests that as the climate warms, California's precipitation mix will shift significantly in favor of rain over snow. That's dire news for our food system, because the Central Valley's vast irrigation networks are geared to channeling the slow, predictable melt of the snowpack into usable water for farms. Water that falls as rain is much harder to capture and bend to the slow-release needs of agriculture.

In short, California's climate, chaotic under normal conditions, is about to get weirder and wilder. Indeed, it's already happening.

What if an 1862-level flood, which is overdue and "more likely than not" to occur with a couple of decades, were to hit present-day California?

Starting in 2008, the USGS set out to answer just that question, launching a project called the ARkStorm (for "atmospheric river 1,000 storm") Scenario. The effort was modeled on a previous USGS push to get a grip on another looming California cataclysm: a massive earthquake along the San Andreas Fault. In 2008, USGS produced the ShakeOut Earthquake Scenario, a "detailed depiction of a hypothetical magnitude 7.8 earthquake." The study "served as the centerpiece of the largest earthquake drill in U.S. history, involving over five thousand emergency responders and the participation of over 5.5 million citizens," the USGS later reported.

That same year, the agency assembled a team of 117 scientists, engineers, public-policy experts, and insurance experts to model what kind of impact a monster storm event would have on modern California.

At the time, Lucy Jones served as the chief scientist for the USGS's Multi Hazards Demonstration Project, which oversaw both projects. A seismologist by training, Jones spent her time studying the devastations of earthquakes and convincing policy makers to invest resources into preparing for them. The ARkStorm project took her aback, she told me. The first thing she and her team did was ask, What's the biggest flood in California we know about? "I'm a fourth-generation Californian who studies disaster risk, and I had never heard of the Great Flood of 1862," she said. "None of us had heard of it," she added—not even the meteorologists knew about what's "by far the biggest disaster ever in California and the whole Southwest" over the past two centuries.

At first, the meteorologists were constrained in modeling a realistic megastorm by a lack of data; solid rainfall-gauge measures go back only a century. But after hearing about the 1862 flood, the ARkStorm team dug into research from Ingram and others for information about megastorms before U.S. statehood and European contact. They were shocked to learn that the previous 1,800 years had about six events that were *more* severe than 1862, along with several more that were roughly of the same magnitude. What they found was that a massive flood is every bit as likely to strike California, and as imminent, as a massive quake.

Even with this information, modeling a massive flood proved more challenging than projecting out a massive earthquake. "We seismologists do this all the time—we create synthetic seismographs," she said. Want to see what a quake reaching 7.8 on the Richter scale would look like along the San Andreas Fault? Easy, she said. Meteorologists, by contrast, are fixated on accurate prediction of near-future events; "creating a synthetic event wasn't something they had ever done." They couldn't just re-create the 1862 event, because most of the information we have about it is piecemeal, from eyewitness accounts and sediment samples.

To get their heads around how to construct a reasonable approximation of a megastorm, the team's meteorologists went looking for well-documented twentieth-century events that could serve as a model. They settled on two: a series of big storms in 1969 that hit Southern California hardest and a 1986 cluster that did the same to the northern part of the state. To create the ARkStorm scenario, they stitched the two together. Doing so gave the researchers a rich and regionally precise trove of data to sketch out a massive Big One storm scenario.

There was one problem: while the fictional ARkStorm is indeed a massive event, it's still significantly smaller than the one that caused the Great Flood of 1862. "Our [hypothetical storm] only had total rain for twenty-five days, while there were forty-five days in 1861 to '62," Jones said. They plunged ahead anyway, for two reasons. One was that they had robust data on the two twentieth-century storm events, giving disaster modelers plenty to work with. The second was that they figured a smaller-than-1862 catastrophe would help build public buy-in, by making the project hard to dismiss as an unrealistic figment of scaremongering bureaucrats.

What they found stunned them—and should stun anyone who relies on California to produce food (not to mention anyone who lives there). The headline number: $725 billion in damage, nearly four times what the USGS's seismology team arrived at for its massive-quake scenario ($200 billion). For comparison, the two most costly natural disasters in modern U.S. history—Hurricane Katrina in 2005 and Harvey in 2017—racked up $166 billion and $129 billion, respectively. The ARkStorm would "flood thousands of square miles of urban and agricultural land, result in thousands of landslides, [and] disrupt lifelines throughout the state for days or weeks," the study reckoned. Altogether, 25 percent of the state's buildings would be damaged.

In their model, twenty-five days of relentless rains overwhelm the Central Valley's flood-control infrastructure. Then large swaths of the northern part of the Central Valley go under as much as twenty feet of water. The southern part, the San Joaquin Valley, gets off lighter; but a miles-wide band of floodwater collects in the lowest-elevation regions, ballooning out to encompass the expanse that was once the Tulare Lake bottom and stretching to the valley's southern extreme. As Jones stressed to me in our conversation, the ARkStorm scenario is a cautious approximation; a megastorm that matches 1862 or its relatively recent antecedents could plausibly bury the entire Central Valley underwater, northern tip to southern.

A twenty-first-century megastorm would fall on a region quite different from gold rush–era California. For one thing, it's much more populous. While the ARkStorm reckoning did not estimate a death toll, it warned of a "substantial loss of life" because "flood depths in some areas could realistically be on the order of 10–20 feet."

Then there's the transformation of farming since then. The 1862 storm drowned an estimated 200,000 head of cattle, about a quarter of the state's entire herd. Today California houses more than 5 million head of beef cattle, of which 70 percent live in the southern part of the Central Valley. The dairy industry adds another 1.7 million cows to the mix; nearly 1.4 million are clustered in the valley. While cattle continue to be an important part of the region's farming mix, they no longer dominate it. Today the valley is increasingly given over to intensive almond, pistachio, and grape plantations, representing billions of dollars of investments in crops that take years to establish, are expected to flourish for decades, and could be wiped out by a flood.

Apart from economic losses, "the evolution of a modern society creates new risks from natural disasters," Jones told me. She cited electric power grids, which didn't exist in mid-nineteenth-century California. A hundred years ago, when electrification was taking off, extended power outages caused inconveniences. Now, loss of electricity can mean death for vulnerable populations (think hospitals, nursing homes, and prisons). Another example is the intensification of farming. When a few hundred thousand cattle roamed the sparsely populated Central Valley in 1861, their drowning posed relatively limited biohazard risks, although, according to one contemporary account, in post-flood Sacramento, there were a "good many drowned hogs and cattle lying around loose in the streets."

Today, however, several million cows are packed into massive feedlots in the southern Central Valley, their waste often concentrated in open-air liquid manure lagoons, ready to be swept away and blended into a fecal slurry. Tulare County alone houses nearly 500,000 dairy cows, with 258 operations holding on average 1,800 cattle each. Mature modern dairy cows are massive creatures, weighing around 1,500 pounds each and standing nearly five feet tall at the front shoulder. Imagine trying to quickly move such beasts by the thousands out of the path of a flood—and the consequences of failing to do so.

A massive flood could severely pollute soil and groundwater in the Central Valley, and not just from rotting livestock carcasses and millions of tons of concentrated manure. In a 2015 paper, a team of USGS researchers tried to sum up the myriad toxic substances that would be stirred up and spread around by massive storms and floods. The cities of 160 years ago could not boast municipal wastewater facilities, which filter pathogens and pollutants in human sewage, nor municipal dumps, which concentrate often-toxic garbage. In the region's teeming twenty-first-century urban areas, those vital sanitation services would become major threats. The report projects that a toxic soup of "petroleum, mercury, asbestos, persistent organic pollutants, molds, and soil-borne or sewage-borne pathogens" would spread across much of the valley, as would concentrated animal manure, fertilizer, pesticides, and other industrial chemicals.

The valley's southernmost county, Kern, is a case study in the region's vulnerabilities. Kern's farmers lead the entire nation in agricultural output by dollar value, annually producing $7 billion worth of foodstuffs like almonds, grapes, citrus, pistachios, and milk. The county houses more than 156,000 dairy cows in facilities averaging 3,200 head each. That frenzy of agricultural production means loads of chemicals on hand; every year, Kern farmers use around 30 million pounds of pesticides, second only to Fresno among California counties. (Altogether, five San Joaquin Valley counties use about half of all the more than 200 million pounds of pesticides applied in California.)

Kern is also one of the nation's most prodigious oil-producing counties. Its vast array of pump jacks, many of them located in farm fields, produce 70 percent of California's entire oil output. It's also home to two large oil refineries. If Kern County were a state, it would be the nation's seventh-leading oil-producing one, churning out twice as much crude as Louisiana. In a massive storm, floodwaters could pick up a substantial amount of highly toxic petroleum and byproducts. Again, in the ARkStorm scenario, Kern County gets hit hard by rain but mostly escapes the worst flooding. The real "Other Big One" might not be so kind, Jones said.

In the end, the USGS team could not estimate the level of damage that will be visited upon the Central Valley's soil and groundwater from a megaflood: too many variables, too many toxins and biohazards that could be sucked into the vortex. They concluded that "flood-related environmental contamination impacts are expected to be the most widespread and substantial in lowland areas of the Central Valley, the Sacramento–San Joaquin River Delta, the San Francisco Bay area, and portions of the greater Los Angeles metroplex."

Jones said the initial reaction to the 2011 release of the ARkStorm report among California's policymakers and emergency managers was skepticism: "Oh, no, that's too big—it's impossible," they would say. "We got lots of traction with the earthquake scenario, and when we did the big flood, nobody wanted to listen to us," she said.

But after years of patiently informing the state's decision makers that such a disaster is just as likely as a megaquake—and likely much more devastating—the word is getting out. She said the ARkStorm message probably helped prepare emergency managers for the severe storms of February 2017. That month, the massive Oroville Dam in the Sierra Nevada foothills very nearly failed, threatening to send a thirty-foot-tall wall of water gushing into the northern Central Valley. As the spillway teetered on the edge of collapse, officials ordered the evacuation of 188,000 people in the communities below. The entire California National Guard was put on notice to mobilize if needed—the first such order since the 1992 Rodney King riots in Los Angeles. Although the dam ultimately held up, the Oroville incident illustrates the challenges of moving hundreds of thousands of people out of harm's way on short notice. The evacuation order "unleashed a flood of its own, sending tens of thousands of cars simultaneously onto undersize roads, creating hours-long backups that left residents wondering if they would get to high ground before floodwaters overtook them," the *Sacramento Bee* reported. Eight hours after the evacuation, highways were still jammed with slow-moving traffic. A California Highway Patrol spokesman summed up the scene for the *Bee*:

> Unprepared citizens who were running out of gas and their vehicles were becoming disabled in the roadway. People were utilizing the shoulder, driving the wrong way. Traffic collisions were occurring. People fearing for their lives, not abiding by the traffic laws. All combined, it created big problems. It ended up pure, mass chaos.

Even so, Jones said the evacuation went as smoothly as could be expected, and likely would have saved thousands of lives if the dam had burst. "But there are some things you can't prepare for." Obviously, getting area residents to safety was the first priority, but animal inhabitants were vulnerable, too. If the dam had burst, she said, "I doubt they would have been able to save cattle."

As the state's ever-strained emergency-service agencies prepare for the Other Big One, there's evidence other agencies are struggling to grapple with the likelihood of a megaflood. In the wake of the 2017 near-disaster at Oroville, state agencies spent more than $1 billion repairing the damaged dam and bolstering it for future storms. Just as work was being completed in fall 2018, the Federal Energy Regulatory Commission assessed the situation and found that a "probable maximum flood"—on the scale of the ARkStorm—would likely overwhelm the dam. FERC called on the state to invest in a "more robust and resilient design" to prevent a future cataclysm. The state's Department of Water Resources responded by launching a "needs assessment" of the dam's safety that's due to wrap up in 2020.

Of course, in a state beset by the increasing threat of wildfires in populated areas as well as earthquakes, funds for disaster preparation are tightly stretched. All in all, Jones said, "we're still much more prepared for a quake than a flood."

3

Pumping Air

The year was 2014, midsummer, the height of California's worst drought in recorded history. Working on a story about the state's nut boom, I headed south from Sacramento into the Tulare Basin of San Joaquin Valley. The farther south I drove down the eighteen-wheeler-choked Highway 5, the more I saw them: dense thickets of almond and pistachio trees, heavy with millions of dollars' worth of nuts, glistening like jewels in the hot sun.

I turned east off the 5, and after a drive through more groves, I arrived in Alpaugh, a farmworker town, population 1,026. A century ago, the landscape surrounding Alpaugh was submerged by the now-vanished Tulare Lake, but not the town itself—it was a high point, an island. These days, it's a grid of low-slung single-family homes with small yards bordered by chain-link fences. Here and there, palm trees hover above the structures. Many yards also contain peach and apricot trees, which appeared to be languishing under heat stress when I visited. The mean annual household income is less than $28,000, and the poverty rate exceeds 40 percent.

I was there to see why Alpaugh's residents, who provide labor for the surrounding nut groves, couldn't drink the water that flowed through their taps—even as the groves continued expanding.

My guide was John Burchard, a soft-spoken, energetic octogenarian who then managed the Alpaugh Community Services District, giving him what had become the Herculean task of delivering safe water to Alpaugh's residents, about a third of whom are children. A lanky man with friendly eyes and a neatly trimmed white goatee, he wore dark jeans, a long-sleeved blue work shirt, and a brown felt cowboy hat turned up sharply at the sides.

Under a morning sun still cranking up to full fire, we stood outside one of the town's two wells, and Burchard explained why Alpaugh's community members were drawing poisoned tap water laced with arsenic at 30 parts per billion—three times the upper limit of 10 parts per billion set by the Environmental Protection Agency.

The well we were looking at was "on the edge of compliance" with the EPA's arsenic limit, Burchard said. "For the last three or four months that we ran this well, it was compliant—just."

Those were the good old days. As the drought-stricken winter of 2013–14 dragged on, Burchard found, the aquifer the well drew from was dropping at a rate of about ten feet per week. By May, it had nearly gone dry. "If we kept using it, we'd soon be pumping air."

So he had to switch to the town's other well, which plunges one hundred feet deeper into the ground. Half a mile away, this well taps from a different part of the aquifer—one, Burchard knew, that was significantly richer in arsenic. With his (barely) compliant well tapped out, Burchard had no choice but to switch to the poisoned one, sending out water with unsafe levels of the chemical.

The fact that the aquifer that sits under Alpaugh had so much arsenic in the first place wasn't a direct consequence of the drought or the ongoing well-drilling boom. Rather, Burchard explained, it was the result of decades of reliance on pumping groundwater to feed agriculture. As the water level dropped, it got closer to Earth's crust, which contains increasing levels of naturally occurring arsenic the deeper you go.

Chronic low-level exposure to arsenic has been linked to heart disease and cancer; children and developing fetuses are particularly vulnerable. So it was not surprising that Alpaugh residents were then—as they still were five years later—relying largely on bottled water for drinking and cooking, spending an average of $400 annually per household—a large burden in a town whose main work source is low-wage farmwork. That totaled about

$400,000 in water expenditures by private citizens. Burchard marveled at the irony. "That's more than our total budget," he said. With an extra $400,000 on top of the $300,000 already at the town's disposal, Burchard could filter the water completely clean of arsenic. "This is a terrible misallocation of resources," he said. With an extra $400,000 on hand, he said, chuckling, "I could send rosewater" through the taps.

Weeks before, then governor Jerry Brown had issued draconian restrictions on water for nonagricultural uses in reaction to the drought. As a result, Burchard had to tell Alpaugh residents they could no longer water their lawns—including vegetable gardens and those fruit trees. This, even though the town drew a tiny fraction of the water from the aquifer, compared with what was drawn by the Alpaugh Irrigation District, which had unrestricted license to pump groundwater to the area's farms.

As Burchard explained what he calls this "bitter irony," I looked at the houses across the street. Peach trees that supplied fresh fruit to cash-strapped farmworker families had to be cut off because of the drought, while the vast nut orchards that surrounded the town, generating what is essentially a luxury product for consumers worldwide, persevered. Any water savings Burchard managed to wring out of the people who employed him would be utterly dwarfed by the water going to farmers. As of 2019, Alpaugh's residents were waiting on the completion of a $3.9 million arsenic-filtration system, paid for with a grant from the state, expected to be ready in late 2020.

Later, as the sun reached its noon-day blitz, we headed out to see the nut boom in full flower. Just a few miles outside town, after making our way down dirt roads alongside irrigation ditches, we encountered a stand of newly planted pistachio saplings. The young trees were arrayed in neat rows extending nearly to the horizon in every direction, with the Coast Range looming to the west: something like four square miles of parched, gray dirt, broken up only by small tree shrubs. Each was attached to a three-foot-high stake to keep it upright and connected by drip-irrigation tubes, about the diameter of a garden hose, that ran along the ground. With the exception of Iowa's seemingly endless cornfields, it was the most impressive example of monoculture—of land used to grow only one type of crop—I'd ever seen.

Four square miles is about 2,500 acres. According to University of California, Davis, estimates, it cost about $2,700 to establish an acre of pistachios in the southern San Joaquin Valley at that time. So we were looking at about a $7 million investment. A mature pistachio grove generates around $3,000 per acre in annual profit. That meant that if all went well, the investment would break even on year eight, the first year it would deliver a full harvest, and then churn out an average of $8 million per year going forward.

Why pistachios and not almonds? Because pistachios can tolerate higher salt content in water than almonds can. Salinization was spreading through the San Joaquin Valley's west side. Facing the problem of diminishing groundwater becoming increasingly concentrated with minerals, farmers throughout the area, including Joe Del Bosque, turned to a less sensitive crop. By 2014, the soil in western Tulare County was already too salty for almonds, making pistachios increasingly the go-to there.

In the depths of the vast orchard, we stopped when we came upon a three-man crew working on a massive well. A red crane attached to a trailer towered some thirty feet into the sky. Beside the trailer, there was a stack of eight-foot metal pipes. Burchard was fascinated, showing the professional interest of a waterman who devotes himself to maintaining wells. We approached to ask what the workers were up to. They shooed us back, pointing to their hard hats to inform us that we couldn't get close without wearing some of our own. Burchard concluded that they were deepening an existing well. As we watched, the men set busily about their task, snatching the pipes with the crane and directing them vertically into the hole, one by one.

Behind the well, a mound of dirt was piled up twenty or so feet into the air. I walked to the top. A reservoir was dug out on the other side, built to hold water pumped up from the well, Burchard explained. The water it held would eventually be siphoned into the drip irrigation tubing and delivered to those hundreds of thousands of young trees.

From my vantage point on the man-made hill, I saw blurry human forms moving toward me in the distance through the knee-high trees. When they got closer, I saw that they were doing some of the only nonautomated work involved in raising nuts: pruning the saplings and making sure they're well attached to the stakes that hold them up. A crew of seven or eight young men were dressed in jeans, long-sleeved shirts, and baseball hats, with cloths covering their necks: protection against the pounding sun. Working in a young orchard is like working in a desert—there's no shade anywhere, no canopy to provide relief. I chatted with them in Spanish as they passed. These days they live in Alpaugh, but they were from Guerrero, a state along central Mexico's Pacific coast. They had no idea who owned this huge tract of land.

From there, we headed back to town to speak to Jim Atwell, general manager of the Alpaugh Irrigation District. He told us the land we'd just seen lay slightly outside his irrigation system—the owners drilled those wells independently, though they drew from the same aquifer as he did for farmers and Burchard did for the town. A fifth-generation farmer from the area himself—Alpaugh was once called Atwell's Island, named for one of his ancestors—he said that people were buying land within his district with the intention of planting pistachios: one hundred acres here, five hundred there, some of which hadn't been farmed at all for years.

I asked him about the huge new orchard we'd just seen. "I used to run a cattle farm out there ten or twelve years ago," he said. Then, the land wasn't irrigated—grass and other scrub sponged up the few inches of winter rain, living long enough to sustain cows during the dry months. Now, if these pistachio trees were ever going to justify the outlay required to get them started, they would require underground water for the next twenty years.

Alpaugh is a microcosm of the state of affairs that persists today in the San Joaquin Valley: toxic water for the low-income workforce, near-unlimited water for nut trees, and a plunging aquifer. What happened? Fitting for an agricultural revolution that began with the gold rush, San Joaquin Valley agriculture is driven by money and power.

When you think of almond farms, the industry would like you to think of independent growers like Joe Del Bosque. I first met Del Bosque in 2014, when I contacted the Almond Board of California for its perspective on the drought. Del Bosque had become a passionate spokesman for the dilemmas independent growers face in a region characterized by a tight labor supply and an even tighter water supply. When we met, he explained how he'd had to fallow hundreds of acres of his prized organic melon fields to divert water to his almond groves, even though the melons require far less water than almonds. But melons are annual crops, so cutting back for a season means a onetime reduction in income; not watering an almond grove could destroy a six-figure investment and a robust multiyear income stream.

Ultimately, farmers like Del Bosque are foot soldiers in California's nut boom, not its leaders. One firm looms over the boom like a colossal marketing display in a supermarket produce aisle. Owned and operated by the Los Angeles tycoons Stewart and Lynda Resnick, the Wonderful Company—formerly known as Paramount Farms and Roll Global—calls itself the "world's largest vertically integrated pistachio and almond grower and processor." Wonderful owns and farms at least fifty thousand acres of nuts in California—more than its closest competitor by a factor of three.

It also buys, processes, and markets nuts from other farmers. Wonderful Pistachio processes an astonishing 80 percent of California's $1.5 billion-per-year pistachio crop, and a substantial but undisclosed share of the $5 billion almond crop. For good measure, Wonderful also dominates the growing, processing, and marketing of two other blockbuster San Joaquin Valley commodities: pomegranates and mandarin oranges (marketed first as Cuties, later as Halos).

The power couple have flexed their political muscles to make sure their plantations get plenty of water from California's publicly funded irrigation projects. They store it for dry years in the Kern Water Bank, a man-made, underground reservoir built by the state of California for public water storage. The thirty-two-square-mile reservoir, located southwest of Bakersfield in Kern County, was meant to collect and store excess surface water flowing from the Sierra Nevada during wet years, banking it for urban areas to use in droughts.

In an infamous 1994 deal known as the Monterey Agreement, the Resnicks' holding company essentially privatized a state-funded water asset designed to be a public good by gaining a controlling stake in the water bank. Ever since, the Kern Water Bank's liquid dividends have accrued to the Resnicks' nut groves, while also giving them the option to sell water to cities if that ever proves more profitable than nuts. Meanwhile, the Resnicks ramped up their campaign donations at the federal level, where lawmakers have a say in how much surface water—which, unlike groundwater, is highly regulated by both the federal and state governments—makes it into the Kern bank.

Having previously turned groundwater extracted from a distant aquifer (Fiji Island) and pomegranate juice (Pom Wonderful) into ubiquitous products—and amassing a $3.8 billion fortune—the Resnicks have more recently focused their attention on pistachios. Consumers nationwide have become accustomed to seeing cardboard "Get Crackin'" displays in the produce section of supermarkets.

Their nut dominance puts the Resnicks at the helm of a multibillion-dollar global market. Saudi Arabia *wishes* it dominated the global oil market the way California does the trade in almonds and pistachios. The state's farms churn out 80 percent of the world's almonds, the great bulk of them in the San Joaquin Valley. Valley farmers are also responsible for more than half the global pistachio output, easily topping their only serious rivals, their counterparts in Iran.

Low in carbohydrates and high in monounsaturated fat, protein, and fiber, these nuts are buoyed by a rising wave of nutritional consensus, diet fashions (gluten-free, vegan, Paleo, low-carb, etc.), and relentless Resnick marketing magic. In the United States, per capita almond consumption jumped by more than 300 percent between 1998 and 2018, reaching 2.3 pounds per year. Pistachios have enjoyed an even wilder ride: Americans were noshing on them at a per capita rate of about 0.2 pounds annually for decades, before consumption abruptly doubled in 2016—when Wonderful's supermarket push began—a level it has held since.

But the main driver comes from abroad. Nearly 70 percent of California's almond crop and nearly half of its pistachios are exported, with China and the European Union as the leading customers. The Indian market, too, is on the move, with annual U.S. almond shipments surging from 77.4 million pounds in 2008 to 193 million pounds in 2017.

The industry got a foothold in Asia by pitching almonds as a go-to healthy snack for the burgeoning, protein-seeking middle class. But research conducted in 2015 indicated that Chinese consumers "were not as receptive to heart health messaging as they were a few years ago, and they did not see heart health as a distinct benefit unique to almonds, since they perceive all nuts this way," the Almond Board of California reported in 2016. So the marketing department switched gears:

> In the minds of Chinese consumers, California has a positive image for its sunny, warm, fresh environment, and easygoing, vibrant lifestyle. Consumers responded positively to a concept following the almond journey through beautiful California and ending in China for consumers to enjoy. The theme "Taste the Sunshine" embodies the California experience and rewards Chinese consumers with a bit of California sunshine with every bite of almonds.

In a 2017 appeal to French women, meanwhile, the pitch centered on almonds as a creation of nature.

The headline reads "*La nature fait bien les choses*," which is an old French proverb meaning "nature does things best." Plants, frogs, hummingbirds, butterflies, and ladybugs are featured in the ad, together with a bowl of almonds, in a natural setting to illustrate nature's admiration for the mighty almond. In other words, all of Mother Nature's creations are perfectly made, just like a healthy, natural almond snack!

The success of these campaigns helped trigger a surge in global demand for the products—and a corresponding spread of almond acreage in the San Joaquin Valley, where production of both is concentrated. Severe recent droughts have not slowed the steady spread of thirsty orchards. Both pistachios and almonds have seen the number of California acres devoted to their cultivation rise without interruption, doubling overall between 2004 and 2017.

As the crops have expanded their domain, the revenue they generate has boomed—albeit with occasional interruptions from external factors, like China's economic slowdown in 2016 and President Trump's trade wars in 2018.

Meanwhile, the Resnicks also use their marketing skills to convince more farmers to plant almonds and pistachios to keep the export markets satisfied. Back in 2015, the Wonderful Company called its pistachio growers to a meeting in Visalia, that early post–gold rush U.S. settlement in the valley, now the seat of Tulare County. The goal was to convince the farmers growing pistachios and almonds for the company to expand their groves.

According to an account in the trade journal *Western Farm Press*, a Wonderful exec got the crowd fired up by playing a clip from *Jerry Maguire*,

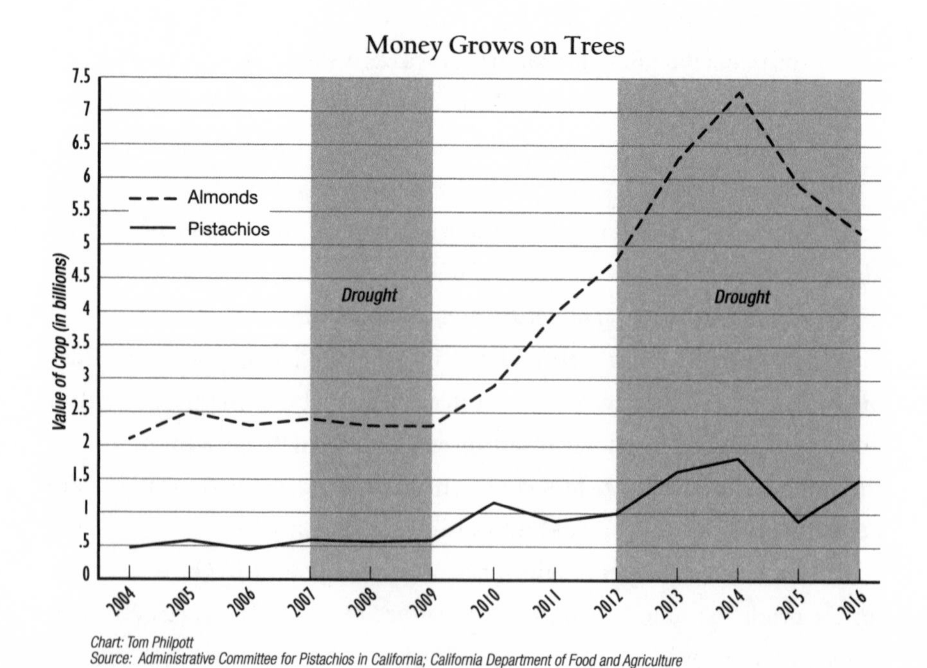

Chart: Tom Philpott
Source: Administrative Committee for Pistachios in California; California Department of Food and Agriculture

the scene in which Tom Cruise thunders, "Show me the money!" The exec then handed the mic to Stewart Resnick himself, who proceeded to, well, show the crowd the money. Resnick told the assembled that once they're up and running, almond groves deliver farmers an average net return of $1,431 per acre—$1.4 million per year for a thousand-acre stand. Pistachios net even more, he added: $3,519 per acre.

But there's a catch, the company acknowledged: the money can flow only so long as water does. According to the *Western Farm Press*, a Wonderful exec revealed to the crowd that the company was spending about $6 million per year on a public relations and lobbying effort to keep water deliveries from the big irrigation projects flowing to the valley; and proposed that the industry come together and commit to devoting a fraction of all pistachio sales to the fund, enough to generate an additional $2 million in a typical year.

"Pistachios are valued at forty thousand dollars an acre. How much are you spending in the political arena to preserve that asset?" the exec asked.

Massive financial interests—banks, pension funds, investment arms of insurance companies—have followed the Resnicks into the nut trade. It's not hard to see why. Putting in a sizable orchard requires deep pockets. Even for a modest two-hundred-acre plot of pistachios, you're looking at upwards of $500,000 to get in the game—and about $1.2 million for almonds, at $6,200 per acre. Then you have to wait three or four years for the orchard to establish and start churning out a harvestable product.

Once you're in, though, the rewards are startling: Davis researchers estimate that once an orchard is up and running, almonds can reasonably generate (depending on yield and price) $1,200 per acre annually above production costs, over a life-span of around twenty years.

So a two-hundred-acre almond farm that cost $1.2 million to establish can throw off $240,000 per year in income. The smart money has sized up the situation and decided to pounce. Wall Street's move into almonds and pistachios is part of a broader trend: the rise of farmland as an "asset class"—finance-speak for a type of investment (e.g., stocks and bonds) that belongs (according to Wall Street salespeople) in every wealthy person's well-balanced portfolio.

The financialization of farming is pitched as a low-risk way for investors to cash in on two long-term trends. The first is a growing global middle class hungry for high-protein, high-status foods like meat and nuts. The second is the steady shrinkage of arable land under pressure from pollution and urban sprawl. Farmland investments are also largely immune from economic shocks, performing well even when stocks and bonds plunge, boosters note. And they offer investors two potential ways to make money simultaneously: from annual income in the form of either crop sales or rent (if it's leased to an independent farmer); and from appreciation, assuming the land's value keeps rising.

Back in the 1990s, U.S. farmland got what every asset class needs to take flight: an index that financial brokers can brandish to clients. That's when a group called the National Council of Real Estate Investment Fiduciaries (NCREIF) launched its Farmland Index, which tracks the annual income and appreciation of U.S. agricultural land held by third-party investors.

Just as the S&P 500 serves as the benchmark for stock investors, tracking and aggregating the share prices of a wide range of publicly traded U.S. stocks, the NCREIF Farmland Index allows investors to assess the money-making performance of farms. The index has given the asset class plenty to boast about over the past decade and a half—values have risen steadily, even when the dot-com bust of the early 2000s and the mortgage meltdown of 2008 tanked stock prices.

Overall, between 1992 and 2017, the NCREIF Farmland Index delivered an average annual return of 11.8 percent, beating the S&P's return of 9.6 percent with much less of the usual stock-market volatility.

Some states—including Midwestern industrial agriculture titans Iowa, Kansas, Minnesota, and the Dakotas—effectively ban ownership of farmland by financial interests. Not California. And the Golden State's nut boom has been a driving force in the success of the farmland asset class. By late 2018, the NCREIF Farmland Index was tracking $9.4 billion worth of U.S.

farm property owned by investors. Of that total, $3.3 billion consisted of "permanent cropland" in California (i.e., nut orchards and wine grapes, another major fixation for institutional investors).

In a 2013 note to clients, Heather Davis, then an executive for TIAA (formerly known as TIAA-CREF), a New York–based retirement and investment fund with $1 trillion in total assets under management, made the case for buying San Joaquin Valley farmland and planting nut groves. Two "countervailing factors"—surging demand and vanishing land—"make the agricultural sector, including almond farms, an attractive long-term investment theme," she wrote. Land suitable for almonds in particular, she argued, has the potential to combine the steady income of bonds with the growth potential of stocks—an investor's holy grail.

TIAA, whose farm assets are now managed by its subsidiary Nuveen, owns 37,000 acres of California farmland, producing "more than 18 million pounds of almonds, enough to circle the world more than nine times or fill the fair territory at Yankee Stadium more than a foot deep," Davis boasted. Following the Resnicks' lead, TIAA not only snaps up land for nut groves but also has pursued a "vertical integration" strategy by buying into the processing and marketing side of the business. In 2011, TIAA bought a minority stake in a large processor called Treehouse California Almonds, "removing an intermediary that was capturing part of the value that we were creating on our farms," Davis explained. Today, TIAA owns a 40 percent stake in Treehouse.

Another financial bigfoot that has invested heavily in this space is the sprawling Canadian insurance and financial services giant Manulife Financial Corporation, through a subsidiary called Hancock Agricultural Investment Group (HAIG). Owning around $3 billion worth of land globally, HAIG calls itself one of the world's "largest managers of farmland investments for institutional investors"—think pension funds, hedge funds, and university endowments.

In a 2018 report to investors, HAIG noted that about 20 percent of its $3 billion portfolio consists of land planted in almonds, and another 18 percent is devoted to pistachios, meaning it owns about $900 million worth of nut land, mostly in the San Joaquin Valley. HAIG owns at least twenty-four thousand acres of almonds, pistachios, and walnuts, making it California's second-largest nut grower, behind only the Resnicks.

Then there's Prudential Financial, the vast insurance and investment conglomerate with assets under management exceeding $1.4 trillion. A Prudential subsidiary called PGIM Real Estate Finance owns a farmland portfolio worth $977 million. The company doesn't divulge its holdings by region or crop type. But in late 2018, its website trumpeted two "recent acquisitions": a pistachio farm in the heart of the San Joaquin Valley that "will provide near term cash flow" to its clients, and an array of land buys in Tulare and surrounding counties that made up a "significant portfolio of well-developed orchards."

In June 2018, PGIM released a brochure titled *Low-Hanging Fruit: Why You Should Plant U.S. Agriculture in Your Institutional Portfolio.* Its goal: to entice people to invest in U.S. farmland. Noting that California boasts one of the globe's five Mediterranean climates, PGIM reminded readers that nuts are mechanically harvested—meaning very low labor costs—and that land planted with permanent crops has delivered an average 14.2 percent annual return over the previous twenty years, trouncing stocks, bonds, the broader real estate market, and timberland, farmland's rival as an alternative investment.

The brochure noted that the financialization of the U.S. farm remains "in its nascent stage." Despite its rapid rise as an investment target, just 3 percent of U.S. farmland is owned by institutional capital, the firm reported. But not for long: "an aging farmer generation, fractional family ownership structure, and technological advances requiring sizable capital investment will naturally transition farmland holdings from individuals to institutions." One topic not mentioned in the pitch: water scarcity in California nut country.

While funds like HAIG and PGIM are tailored for big institutional investors, anyone can buy shares in Gladstone Land Corporation, a publicly traded real-estate investment trust that owns sixty-three thousand acres in six states, including more than five thousand acres of pistachios and almonds in California. "Our strategy is to own farms that produce healthy foods that are found in the produce and nut sections of your grocery stores," the company's CEO, David Gladstone, stated after buying two San Joaquin Valley orchards for a total of $6.9 million in 2018. Major holders of Gladstone Land stock include Wall Street titans Blackrock, Wells Fargo, Morgan Stanley, and Deutsche Bank.

The gusher of institutional cash flowing into the San Joaquin Valley is unlikely to stop anytime soon. In a 2018 report, HAIG took the measure of what it called the "investable universe of farmland"—that is, land not currently owned by investment firms but ripe for the plucking. The group scanned the globe for "core investment geographies that offer a relatively secure business environment" and "the necessary scale for institutional farm management." The group identified 570,000 acres of San Joaquin Valley land suitable for almond investments, and another 50,000 acres for pistachios. (Three other California crops made the list: grapes, oranges, and walnuts, at 620,000 acres, 450,000 acres, and 123,000 acres, respectively.)

Again, just a few years after an epochal drought and amid mounting research showing that San Joaquin Valley agriculture is far overstepping its water limits, the topic of water availability is missing from the report.

Just a half-hour drive into the Coast Ranges from southwestern Kern County, in the southern extreme of the San Joaquin Valley, lies the gorgeously austere Cuyama Valley, a narrow flatland cut into the mountains by the ephemeral Cuyama River, which runs dry during the region's long, scorching summer. Even if you've never heard of it, you've almost certainly eaten carrots grown there. The Cuyama Valley is one of the four main nodes of California carrot production—and California, in turn, is the source of more than 80 percent of U.S.-grown carrots. Think of it as an appendage of the San Joaquin Valley devoted to that crunchy, vitamin-packed orange root vegetable.

While the San Joaquin Valley can at least hope for annual infusions of surface water from the state and federal water projects, the Cuyama is a water orphan. It relies 100 percent on ancient underground aquifers. But that hasn't stopped enormous agribusiness players from setting up shop there.

On the dusty valley's eastern half, vegetable giants Bolthouse and Grimmway—which together sell 80 percent of U.S.-grown carrots—make the desert bloom with vast sheets of green, watered by wells that have been pulling from underground aquifers for so long that the region has landed on California's list of "critically over-drafted basins." On the western edge, where the aquifers lie deep underground and contain less water, desert scrub and sage hold sway. Until 2014, cattle ranches and the odd small vineyard were the only signs of agriculture.

That year, Harvard University's $36 billion endowment, the largest of any U.S. university, spent $10.1 million for an unirrigated 7,600-acre cattle ranch in the midst of the state's worst drought on record. Its intention: to install a vast vineyard, irrigated with groundwater.

On a June 2017 morning under an already-blistering sun, Steve Gliessman explained to me that wine grapes can actually grow there with little or no extra water. Condor's Hope, the five-acre organic operation Gliessman runs with his wife, Roberta Jaffe, lies in the valley's west side, in the foothills above Harvard's land. "There's just not that much water down there," Jaffe said, pointing down. "That's why the carrot people haven't come here." Most of the land was devoted to sparsely stocked cows, with pasture that relied on an annual average of five inches of rain—terrain known in farming circles as "marginal."

Condor's Hope looked like a farm you might find in the arid southern Mediterranean—a small spread of grapes and olive trees ringed by scrubby pastures, with a stunning mountain silhouette looming behind. Tall, sixty-something, and sporting a gray mustache and a broad-brimmed hat, Gliessman dropped to his knees and ran his hands through the dirt. It appeared way too parched and dusty to support a weed, much less the vines on his five-acre organic vineyard, each looking like a skinny-trunked bonsai tree. Yet their foliage was remarkably green and perky in the unrelenting heat. Each plant held clumps of small green grapes, a white variety adapted to hot and arid conditions.

Growing wine grapes here is tricky, said Gliessman, who moonlights as a farmer when he's not teaching agroecology at the University of California, Santa Cruz. Summers are punishingly hot and dry; deceptively mild winters can give way to spring frosts that hit just as the vines bud, killing the nascent fruit and destroying the year's vintage. Making their task even trickier, Gliessman and Jaffe "dry farm"—grow their grapes with virtually no added irrigation.

Gliessman noted that nearly all of the five inches of rain that fall in the valley come in the winter. When the rains have passed, he tills the soil. The surface quickly dries into a three-inch dry layer of "dust mulch" that acts as a barrier, holding in moisture below and insulating it from heat, which would cause evaporation. This meager supply of water forces the vines to develop roots that plunge as deep as forty feet into the subsoil—turning them into tiny wells that draw from underground aquifers. Using these techniques, the farm had been churning out small quantities of well-regarded wines for twenty years, reverting to light early-season irrigation only in years of extreme drought.

While Gliessman and Jaffe had cracked the code of dry-farming grapes in the desert, what they couldn't figure out was why the secretive managers of Harvard University's endowment made that massive investment in 2014. Or why, rather than keeping it in cattle, like the great bulk of surrounding land, the group opted to develop what emerged as the largest vineyard for miles around.

Already, nearly one thousand acres of the Harvard holding had been planted in vines—two hundred times the size of Gliessman and Jaffe's operation. The couple saw it every time they entered or exited their farm. On the way down from the Sierra foothills from Condor's Hope, the Harvard vineyard rose like a vast oasis amid the desert scrub: long rows of metal stakes spaced a few feet apart, stretching nearly to the mountain ranges on either side of Highway 166, the thoroughfare that traverses the valley, for about a mile. In some parts of this sea of gleaming stakes, small vines crept up; in others, crews of workers braved the sun, planting saplings.

Unlike Condor's Hope's vines, Harvard's wouldn't send roots deep into the earth in search of water. Dry farming high-quality wines is a fussy business and really works only on a small scale. Harvard was developing its vineyard in the conventional way, counting on a plentiful supply of irrigation water. It had dug twelve deep wells and set up drip irrigation lines to keep the plants watered. Harvard's investment arm maintains a strict code of silence with the media, so it's impossible to tell exactly how much it invested. Establishing a vineyard isn't cheap. Land preparation, stakes, trellises, drip lines—it all costs about $15,000 per acre. That means, in addition to its initial $10.1 million purchase price, Harvard invested something like $15 million in the project—and that's not counting those dozen wells.

It's not just the meager water table in the area that made the project puzzling. Just months after Harvard closed its Cuyama deal, the California legislature reversed a long tradition of Wild West–style water management and passed a landmark groundwater-preservation law. By 2040, it could drastically curtail the amount of water the vineyard can draw from the ground—in turn slashing the value of the investment. Why plunge millions of dollars into a water-intensive project in a water-stressed region that will soon be subject to pumping restrictions? More mysteriously still, the Condor's Hope farmers and other locals said, the Harvard team had at least publicly stayed out of the local process that would ultimately decide pumping guidelines.

Back in 1997, the Harvard Management Company, the division of the university that oversees the endowment, charted a bold new course. Instead of buying, say, shares in a paper company, they famously bought a timber forest in Appalachia. To this day, most university endowments avoid "real assets," on the theory that managing them is risky and that they're difficult to unload during a slump. But Harvard pressed on, diversifying into timber plantations in Brazil and Central America and massive dairy operations in New Zealand. Eventually, its natural resources portfolio ballooned to $4 billion.

The fund's move into California wine was part of that trend. It started in 2012—not in the Cuyama Valley, which has no reputation in the wine world, but rather about a hundred miles to the northwest, in the Paso Robles area, the wine region that served as the backdrop for the 2004 oenophile comedy *Sideways* (and has only gotten more fashionable since).

In Paso Robles, the endowment played the heavy in water politics, which were particularly fraught at the time because of the drought: in 2013, as the water scarcity took hold, the group conveniently dug several wells just before the county imposed a well moratorium in response to plunging water tables. And Harvard's representative on the ground pushed hard (though ultimately unsuccessfully) to dominate the governing board that will decide how much water can be pumped out of the region's aquifers as California's groundwater legislation takes shape.

But by 2016, the endowment had plunged into crisis. The natural resources division, once the jewel of the crown, had begun to underperform—hammered by an overall drop in agricultural commodities—and the Harvard Management Company, which had for years consistently generated higher returns than its Ivy League peers, emerged as a laggard.

That year, the fund poached N. P. Narvekar from the endowment of Columbia University, which shies away from the direct investments in productive land favored by Harvard. Under Narvekar, Harvard Management Company changed directions: it laid off half its staff—including prominent managers of its natural resources portfolio—and announced in a press statement that it would begin "refining HMC's natural resources strategy and streamlining assets." By June 2017, it had sold its massive dairy operations in New Zealand and hinted that more sales could be in the works.

What the new direction means for the fund's California wine holding is anyone's guess. But as of 2019, workers were still scrambling in the Cuyama heat to maintain the desert vineyard.

Meanwhile, the water situation is dire. According to a 2015 USGS assessment, growers in the eastern part of Cuyama Valley were pumping water out of the ground at a rate twice as high as natural replenishment—even before the drought that began in 2011. The new state groundwater law mandates that such "critically overdrafted basins" form locally controlled sustainable groundwater agencies that must come up with a plan to bring their aquifers into balance by 2040—meaning, theoretically at least, water can be pumped out only at the same rate that it naturally replenishes.

The massive carrot interests that dominate agriculture in the valley have taken a decisive approach to protecting their land investments as groundwater restrictions come into focus. Led by Bolthouse and Grimmway, they formed the Cuyama Valley Basin Water District soon after the groundwater legislation passed, and that group will have five members, representing about 44 percent of the vote, on the valley's Groundwater Sustainability Agency. Surprisingly, Harvard did not join the water district and has no apparent representative on the GSA. The Cuyama Valley Basin Water District even made a push, ultimately unsuccessful, to draw the GSA boundary such that Harvard's vineyard does not fall within it. So the fate of Harvard's investment might lie in the hands of other Big Ag interests that will be vying for the same vanishing pool of water that Harvard's vines need.

The groundwater legislation may already be pushing down the value of the investment. In central Kern County, the market price of land that relies on groundwater for irrigation, like Harvard's Cuyama vineyard, fell by about 28 percent between spring 2015 and spring 2019. Land with access to surface water, by contrast, held its value over the same period.

Since Harvard bought the Cuyama land several months before the groundwater act passed, perhaps it was betting that California's Old West groundwater rules would survive the drought. But when I spoke to Michael Ming, a veteran appraiser for Bakersfield-based Alliance Ag Services, he said that as early as 2013 "there were whispers that something had to give"—the drought was so severe, and groundwater was disappearing so rapidly, that even ag interests knew that they couldn't block legislation forever.

When he first heard about the Harvard deal in 2014, Ming was "shocked," he said. "I was like, *what*?" It didn't make sense to him that a savvy investment fund like the Harvard endowment would drop big bucks to establish a large vineyard in a marginal region with a stressed water table while rumors swirled of coming water-pumping restrictions. That it had since plunged ahead with pricey vineyard development only compounded his puzzlement, he said.

Like their counterparts in California's other ag-heavy, water-poor regions, Cuyama's Sustainable Groundwater Agency faces a tough and contentious task. Big Carrot and Harvard are drawing from aquifers that also provide water for New Cuyama, a town with a population of about a thousand, as well as regional elementary and high schools. Unlike Harvard, the community does have a seat at the GSA table—Paul Chounet, director of the Cuyama Community Services District (and the outgoing principal of its high school) is on the board.

Today, the area's declining water table means the town well is pumping from deep in the aquifer, where naturally occurring chemicals are concentrated, including arsenic, the same chemical that causes trouble in Alpaugh. Chounet says the town's filtration system keeps the water up to California safety standards. Keeping it that way is pricey—minimum water bills for residences are around $160 per month. Despite the high costs, many locals aren't convinced. Several people told me that most residents avoid drinking from their taps and buy bottled water. Megan Harrington, a server at the restaurant in the town's lone hotel, explained why, echoing others I spoke to throughout the community. "The water is horrible," she said. It tasted and smelled of chemicals, and she used it neither in cooking nor to bathe her infant daughter. She resented having to spend $30 per month on bottled water on top of her water bill.

Meanwhile, the carrot interests are settling in to defend their ability to draw water. The lawyer for the Cuyama Valley Basin Water District—the hastily formed body representing established growers in the area, but not Harvard—is Ernest Conant. Ming called Conant the "go-to guy for a lot of big players and water districts." Conant represented the Kern Water Bank, the water horde owned the Resnicks.

As of late 2019, almond and pistachio plantations were still expanding at a steady clip, with no sign of slowing over water-availability concerns. The ongoing influx of big capital into the valley is a signal that Wall Street thinks that well-capitalized and -connected nut and wine operations can keep the water flowing even as the Sierra Nevada snowpack declines, the valley floor sinks, aquifers empty, and water gets scarcer and more poisonous for residents. A bet, in short, that as water grows more scarce, its tendency to flow toward money only increases. While people making these wagers might be acting rationally in terms of their own financial interests—squeeze as much profit out of the land while the getting's still good—the number one source of U.S. fruits and vegetables faces severe pressure. As Joe Del Bosque's story shows, dwindling water means ever more emphasis on pricey export-oriented snack crops—and less on fruit and vegetable crops. And if climate models about the future of California weather hold true, megadroughts and megafloods will add severe chaos to this race to the bottom of the aquifer.

4

Empire of Dirt

The first time I set foot on Iowa's famously rich soil, I expected a quaint rural landscape, given that farmland stretches over more than 90 percent of the state's territory. But my points of reference misled me. I had lived on a tiny farm amid the Appalachian Mountains of North Carolina for four years; the ten years before that, I'd lived in Mexico City and New York City. I didn't know from vast fields and wide horizons.

Instead, I found myself barreling down the highway, surrounded on all sides by seas of green corn and soybean plants. As I passed row after row after row, marking my progress by that little hypnotic line that forms as your eye picks up the gap between them, I thought, What a triumph of high technology.

White settlers fleeing the already overexploited East Coast showed up in what is now Iowa in the 1850s and found marshlands and prairie, with hundreds of species of perennial wild grasses and legumes. Flowers towered over the newcomers' heads, their roots plunging just as deep into the earth, burying carbon snatched from the atmosphere. Vast herds of bison ate their way through meadows, stimulating new plant growth and recycling nutrients through their manure. Native American peoples played an active role in managing the ecosystem; they periodically set fires that quickly freed nutrients and prevented trees from establishing, allowing prairie plants to thrive without being shaded out. These interactions among people, plants, animals, and climate left behind a majestic store of loamy, fertile topsoil.

From that to *this* in 150 years: trillions upon trillions of plants from just two species, packed as close together as city commuters on the subway at rush hour, but much more orderly. Just 0.1 percent of Iowa's tallgrass prairie remains today. As for that gorgeous cache of topsoil that was fed by and supported the prairie, the state has surrendered fully one half—and counting—of it, on average, since being subjected to the plow.

In California, American capitalism has turned a region with robust water resources and a Mediterranean climate into essentially a massive fruit-and-vegetable factory—and pushed this rare ecosystem well beyond its natural limits. As the state's spectacular droughts, floods, and fires stalk one part of our food supply, a more subtle but equally devastating ecological crisis has taken root in our other key node of food production, the Midwest.

Touch down in the region on a midsummer day, and what you see will look nothing like an ecosystem in a state of unraveling. Rather, you'll be confronted with the spectacle of almost unfathomable bounty: on a summer day now, a vast carpet of corn and soybeans stretches to the horizon in all directions. If you look closely through the seas of bright green (corn) and deep green (soybeans), you might see off-color misfits poking up here or there: weeds that have evolved to resist the annual cascade of herbicides. Otherwise, plant species that aren't corn or soybeans are rare, unless you count the forlorn prairie-grass remnants that line the roadside, awaiting the mower's blade.

On my first visit to this landscape, these fields struck me as the most impressive and brutal example I'd seen of humanity's will to reorder landscapes to its whims—as impressive as the Manhattan skyline, or the view of the valley occupied by Mexico City that you get from an airplane.

The degree of industrial agriculture's triumph over the land defies belief. Nearly every square foot of ground between towns and cities is governed by technical wizardry. The seeds for these crops contain genomes that have been tweaked and patented by some of the globe's largest chemical companies, which also sell the poisons that engage weeds, bugs, and fungal pathogens in an endless battle between chemical engineering and evolution. Great air-conditioned machines called combines roam the land; from their lofty heights a single person can manipulate thousands of acres with minimal physical effort.

Even the hogs are highly regimented. Hogs once served as a linchpin of diversified farms, recycling farm waste into high-value meat and soil-feeding manure; today they are industrial cogs, stuffed together by the thousands in factorylike enclosures, designed to process overabundant corn and soybeans (goosed with growth-enhancing pharmaceuticals) into bacon and burrito filling. Hogs raised at industrial scale generate many times more manure than nearby land can process, turning what should be a soil-building resource into an engine for air and water pollution. Find yourself within a few hundred feet of one of these operations, known as concentrated animal feeding operations (CAFOs), and you'll smell it: an overwhelming putrid stench that triggers a flight response. *Get out*, your body seems to yell. (And for good reason: CAFOs emit airborne toxins known to cause respiratory disease.)

The sheer efficiency (to say nothing of concentrated hog production's potent stench) drains the landscape of people. Pesticides, synthetic and mined fertilizers, and large machines replace work once done by bodies, brains, and biodiversity in the field.

These days, the intensification continues apace. Despite five years of low prices from overproduction, Iowa farmers churned out their highest-ever soybean crop and third-largest corn crop in 2018—and planned to boost production in 2019 before severe spring flooding interfered with planting. In spring 2020, despite yet another drop in prices spurred by the COVID-19 crisis, farmers were gearing up to devote more land than ever to corn and soybeans, the U.S. Department of Agriculture reported. In 2018, Iowa's hog population swelled to 23.6 million—4 percent higher than the previous year and more than 30 percent of the entire U.S. hog herd. Hogs now outnumber the state's people by a factor of nearly eight.

Iowa sits at the center of the Corn Belt, a cluster of states in the northern Midwest, where nearly 90 percent of U.S. corn and 80 percent of soybeans are grown. These are by far the nation's two most prolific crops; in the Corn Belt, they blanket a land mass roughly equal to 1.5 times the total area occupied by California. All told, U.S. farmers produce around 35 percent of the world's corn and 34 percent of its soybeans, making the Corn Belt a key node in the global food system.

To millions of Americans, the Corn Belt exists as a vast and vague hinterland. But its products are a weight-bearing pillar of our material existence. The corn and soybeans grown there course through the U.S. food system—and our bodies. You've probably never directly eaten the region's big two commodities: "dent" corn is distinct from the sweet corn you relish from the cob, and the soybeans are entirely different from edamame. The region's industrially grown corn and soybeans are the food system's blood—unseen but fundamental. Until we come up with a better way to sustain the system (a crucial challenge for many reasons), these crops are a central fact of our existence.

Corn and soybeans are the basis for the record-breaking U.S. meat habit, the feed behind nearly every chicken breast, bacon slice, and burger we eat. For more than a generation, Americans have been accustomed to consuming more than a half pound of chicken, pork, and beef daily—a birthright enjoyed in few places on Earth, ever. Meat, dairy, and eggs—the great bulk of it from animals fed with corn and soybeans—supply 26 percent of calories in the average American's diet.

In addition to providing the building blocks of industrially produced meat, corn and soybeans are the raw material par excellence for the packaged and fast-food industries: key providers of its fats, sugars, thickeners, and flavoring agents. Soybean oil alone owns 57 percent of the U.S. vegetable oil market, and is so ubiquitous in the food system that it accounts for 7 percent of the total calories consumed by Americans, as the Soy Nutrition Institute boasts. Corn owns another 11 percent of the vegetable oil market—and thus provides an additional 1 percent of calories. Look for soybean and corn oil on labels, and you'll see them everywhere. Likewise high-fructose corn syrup (HFCS). Its star has dimmed in recent years, but HFCS and other corn-derived sugars (with less demonized names, like dextrose) make up nearly 45 percent of the U.S. sweetener market, delivering another 7 percent of the calories consumed by Americans. Byproducts like cornstarch and soy lecithin are widely used in processed foods as emulsifiers and thickeners. All told, corn and soybean and their derived products, animal and otherwise, account for more than 40 percent of the calories in the average American's diet.

And the Corn Belt's commodities don't just provide a massive portion of our sustenance; they also help move us around, literally. These days, corn is so abundant that about a third of the U.S. corn crop is transformed into ethanol, accounting for more than 10 percent of the blended gasoline we burn in our cars. Soy-derived biodiesel has a growing share of the much smaller U.S. diesel market.

But serving as the pillar of a food system that feeds a nation of 350 million people, and providing a tenth of the gasoline for a society that relies on gas-guzzling cars for transportation, doesn't exhaust the Corn Belt's bounty. More than 50 percent of U.S. soybeans, 25 percent of pork, and about 13 percent of corn are exported, mostly to satisfy the rising demand for meat among middle-class and wealthy consumers in Canada, Europe, and fast-developing countries like China.

This extraordinary industrial agglomeration is sustained at an even more extraordinary cost. Over the course of my trips to Iowa, I saw firsthand how growing the same two crops year after year over a massive swath of land destroys soil and fouls water—undermining the very ecological basis of food production and human habitation in one of the globe's great breadbaskets. The quality of a diet based on derivatives of corn and soybeans is debatable, to say the least. But setting that problem aside for now, a different question comes into focus: Who profits from this massive bounty?

Among some environmentalists and libertarians, it's easy to paint big-time corn and soy farmers of the Midwest as villains: fat cats harvesting big bucks from on high in their gigantic combines while wantonly spraying chemicals. Yet even with federal crop subsidies and government-backed insurance, large-scale farming in the Corn Belt is a pretty awful business, and it responds to market forces differently than other enterprises within modern capitalism do.

When the price of computer chips falls, for example, dominant chipmakers Intel, AMD, and Nvidia have a simple remedy: they make fewer chips. With a dearth of chips on the market, device makers eventually begin to bid the price back up, so chipmakers respond by boosting production, and the whole cycle begins anew.

But computer chips and crops function differently. Say you're an Iowa corn farmer and the price of corn drops after you've planted the spring crop. Unlike Intel, you can't slash production anytime soon; you have to wait until the next season's planting.

Worse still, when the time comes to put the next season's crops into the field, you're faced with a harsh fact. If you decide to plant less corn, there's no guarantee that the corn price will rise. Why? Because unlike Intel—which essentially shares the chip market with just two other firms—you have thousands of competitors. Unless you can figure out a way to organize a significant portion of them to join you in cutting production, you're not going to succeed in pushing prices up.

Since no mechanism exists to coordinate farmers in their planting decisions, they tend to respond to low prices in a way that would be alien to an Intel exec: they plant more corn. In their individual calculations, if they're going to hold their income steady while prices fall, they'll have to bring more product onto the market, to make up in volume what they're losing in price. But since thousands of other farmers are making the same decision, the market just gets flooded with corn, and prices fall further.

Throughout the Corn Belt, these economic forces have put farmers under massive pressure, helping clear the countryside of people and leading to ever fewer and larger farms. In 1940, Iowa supported 213,000 farms. By 2007, the state's total number of farms had dropped to 92,856. The decade after saw an ethanol-fueled hike in corn prices followed by a prolonged bust. Iowa lost nearly 7,000 more farms, settling at 86,000 in 2018, representing an erasure of 60 percent of the state's farms since the dawn of industrial agriculture in the 1940s. Surviving farms exploded in size. Between 1982 and 1997, the median Iowa farm surged from 395 acres to 869 acres, and has continued rising since.

But there's another, longer-term problem that haunts Corn Belt farming. To put it in economists' jargon, productivity outruns demand. Simply put, farmers in the post–World War II years—and the seed, chemical, biotechnology, and heavy-machinery industries that supply them—kept figuring out new ways to squeeze more and more food out of less and less land, but the human body's caloric needs don't change much. Food demand, economists would say, is pretty inelastic: each person can only eat so much.

Between 1948 and 2002, total U.S. agricultural output rose by a factor of 2.6, and has held steady since, while population rose by a factor of 1.96. It's no wonder that the prices farmers fetch for their goods have tended to steadily fall in real terms.

A 2003 paper authored by a group of University of Tennessee agricultural economists led by Daryll Ray demonstrates how this works. Farmers are quick to adopt technologies (think high-powered tractors, pesticides) that promise to reduce their costs and help them produce more with less labor. But since everyone embraces the same technologies, prices of agricultural goods drop. The lower prices then encourage the adoption of more cost-reducing technologies, and prices continue their slide. In other words, farmers exist on a technology treadmill, and farmers face "constant price pressure, with periods of brief reprieve generally the result of disasters or other random events," as Ray and his team put it.

Between 2006 and 2012, for example, corn and soybean prices surged, bolstered by growing ethanol demand and speculative bets from Wall Street investors fleeing the real estate busts. Of course, those external factors, and not farmers' own planting decisions, sparked the rally. Moreover, farmers quickly responded to the windfall by scrambling to plant more corn and increase yields per acre, hoping to take advantage of high prices—and predictably enough, prices eventually tumbled, and have been low ever since.

Finally, while crop prices are weighed down by the hypercompetition of commodity markets, farmers' cost of doing business tends to work the opposite way. When prices and revenue rise, farmers' costs—machinery, seeds, chemicals, land—rapidly rise in response, limiting profit. But when prices and revenue drop, costs do not fall quite as rapidly, leading to losses.

Ultimately, the two sides of the equation balance each other out. "Agricultural returns tend to be cyclical in nature, a few years of good returns followed by a few years of negative returns," the Iowa State University economist Chad Hart wrote in 2016. Farming is the ultimate competitive industry, and "the long run profitability of a competitive industry is zero."

These charts, drawn from U.S. Department of Agriculture figures for corn and soybean costs and returns, drive home Hart's point.

During the Great Depression of the 1930s, the Roosevelt administration developed programs to help farmers manage planting decisions and maintain profitable crop prices. But starting in the early 1970s, government policy slowly shifted into a backstop to keep ever-larger farms in business while encouraging maximum production. Between 1995 and 2017, while corn and soybean profitability lurched between boom and bust cycles, federal government programs paid Iowa farmers more than $26 billion in

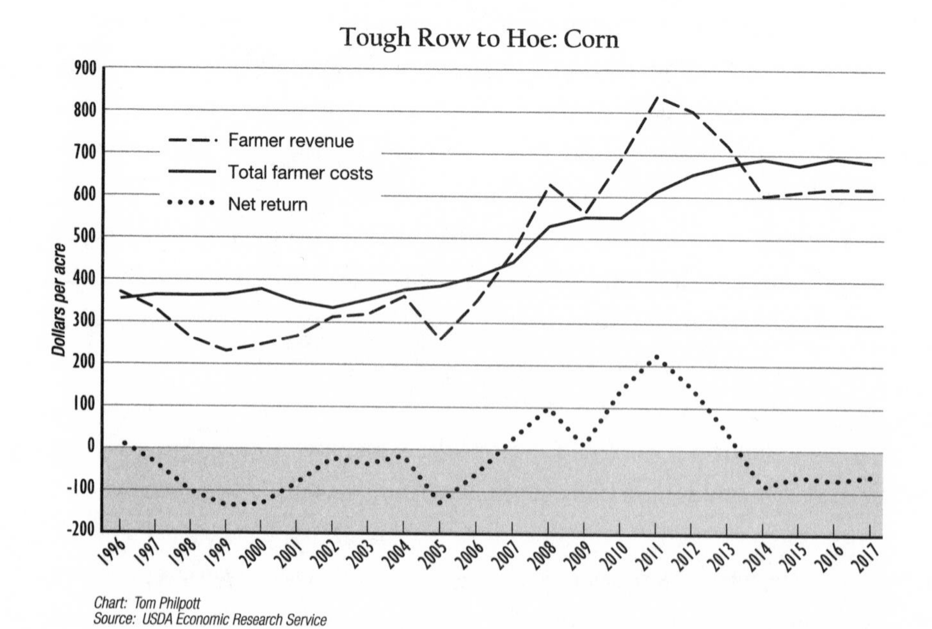

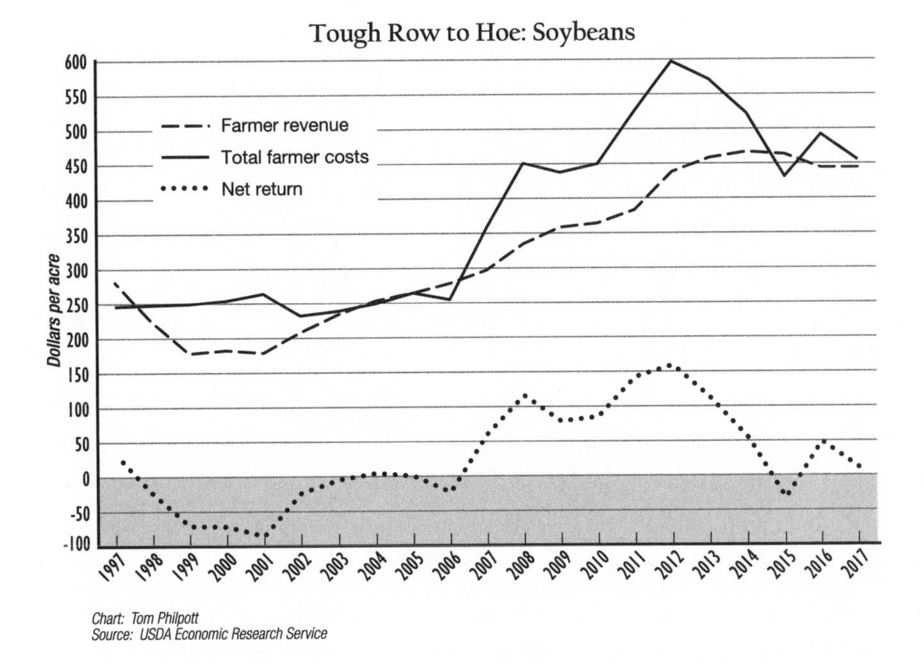

Chart: Tom Philpott
Source: USDA Economic Research Service

commodity and crop-insurance subsidies. The great bulk of that went to corn and soybean growers, the majority of them clustered in the Corn Belt states. All told, those growers received nearly $149 billion in federal subsidies between 1995 and 2017—the lifeblood that sustained a long-run profitless enterprise.

If farming corn and soybeans is such an economically marginal activity—one that annually requires billions in government support to maintain—that means the farmers themselves aren't the ones reaping the economic benefits. Who cashes in on the bounty? In short, the Corn Belt's farmers are caught between a handful of seed, pesticide, and fertilizer firms, known as input suppliers; and a few enormous buyers, in the form of grain trading firms and meatpackers. The input suppliers occupy a position not unlike makers of mining equipment during a gold rush; the buyers thrive by transforming all that cheap bounty into profitable meat, cooking oil, sweetener, and car fuel (ethanol). The degree to which farmers deal with oligopolies—industries dominated by a few firms—at virtually every turn is startling.

Let's start with farm equipment, the high-tech combines and tractors farmers use to till, plant, harvest, spray, and generally mechanize the hard labor of running a corn and soybean operation. As recently as the 1950s, farmers could choose from no fewer than eight companies when buying these big-ticket machines. Today there are three, one of which is dominant.

Combines—short for "combination harvesters"—are the massive, diesel-sucking machines required to hoover up the vast stands of corn and soybeans that characterize a modern Midwestern farm. John Deere alone sells 63 percent of the combines bought in the United States; the sprawling conglomerates CNH Industrial and AGCO vie for the rest.

Farmers have more options when they're shopping for tractors, which are needed to pull seeders, tillers, fertilizer applicators, and pesticide sprayers. But only just. Deere has 53 percent of the tractor market; CNH has 35 percent. Several other smaller firms—including AGCO and Kubota, which focuses on tractors for smaller-scale, more-diversified farms—divvy up the remaining 12 percent not controlled by the two biggest players.

A new combine appropriate for running a typical Iowa farm costs north of $400,000; a new tractor starts at more than $150,000. As the market in heavy farm machinery has consolidated, the machines themselves have gotten bigger and more powerful—and more expensive to keep up. An Iowa State University survey found that in 2007, the average Iowa farm had a combine with an engine rated at 230 horsepower. By 2015, the average had leapt by nearly 50 percent, to 342 horsepower. That addition of power led to a nearly 30 percent leap in harvest capacity per run. But over the same period, farmers' repair costs per acre jumped, too—from $3.33 to $5.24, a nearly 60 percent increase.

One reason for the rise in repair costs is that John Deere and its few peers in the big tractor and combine business use their market heft to control these massive machines even after they've left the lot. For generations after the advent of the mechanical tractor, farmers could tinker with and fix their own machines, enjoying ready access to replacement parts and repair manuals. These days tractors are like enormous rolling smartphones—controlled by interlinking, proprietary software that's digitally locked to prevent user modifications. Draconian warranties also forbid farmers from repairing their rigs, shunting them to authorized service providers, where even minor fixes can cost thousands of dollars and be quite time-consuming—devastating when a farmer needs to harvest.

Farmers throughout the Corn Belt have organized to promote "right to repair" bills, which would force the manufacturers to make manuals and repair-related resources available to the public, not just their own affiliated dealers and service shops. Big Tractor, in league with Apple, Tesla, and other corporations that want to capture the profits from repair, have lobbied vigorously and funded campaigns to defeat them.

Not surprisingly for a company that sells such expensive equipment to such an economically stressed customer base, John Deere offers financing and leasing options. Here's the financing pitch on John Deere's website:

One simple loan—Conveniently package equipment, implements, extended warranties and attachments with one rate and one statement.

Flexible payments—Choose between fixed and variable rates and pick your payment date.

Support tailored to your business—We can adjust your payment schedule according to your unique situation.

John Deere's financial-services division ramped up after 2013, when corn and soybean prices dropped after several years of being puffed up by the U.S. ethanol boom. The company's annual revenues from farm-related equipment sales fell from $29 billion in 2013 to $23 billion in 2018, as low crop prices discouraged farmers from making large purchases. Over that same period, revenues from its financial services operations jumped from $2.5 billion to $3.5 billion.

John Deere's financial arm not only fronts money to farmers to buy combines and tractors; it also provides "short-term credit for crop supplies such as seeds, chemicals and fertilizer, making it the No. 5 agricultural lender behind banks Wells Fargo, Rabobank, Bank of the West and Bank of America," the *Wall Street Journal* reported in 2017. In 2013, the financial arm delivered less than a fifth of the company's profit; by 2016, it was contributing nearly half.

Then there are "inputs," the consumable stuff farmers need to produce their crops: fertilizers, seeds, and pesticides.

Fertilizers, which farmers usually inject into their fields (with the help of one of those pricey tractors) before planting, are typically broken down into three "macronutrients" vital to plant growth: nitrogen, phosphorus (in the form of phosphate), and potassium (as potash), known by their periodic-table initials, NPK. In chapter 6 we'll get to the spectacular environmental damage these chemicals rack up when they're overused. For now, I'll focus on the payout.

Four massive companies loom over the $11 billion U.S. fertilizer market. Spawn of the 2018 megamerger between Canadian giants Potash Corporation of Saskatchewan and Agrium, Nutrien alone churns out 63 percent of the potash produced in the United States, about a fifth of the nitrogen, and 25 percent of the phosphate.

Nutrien essentially shares the U.S. potash and phosphate markets with just one other player, Mosaic, a spin-off of the agribusiness giant Cargill. Mosaic owns the other 47 percent of the U.S. potash market; and in phosphate fertilizers, Mosaic trumps Nutrien's share, with about 60 percent of U.S. phosphate fertilizer capacity, through its massive phosphate-rock-mining operations in Florida.

According to the investment-analysis firm Morningstar, Mosaic's position in the global phosphate market is so dominant that it can affect prices farmers pay for fertilizer by cutting production. "In recent years, the company has reduced its phosphate production to support higher global prices as more low-cost supply has come online from Morocco and the Middle East," the Morningstar analyst Seth Goldstein wrote in March 2019.

"Doing so maximizes phosphate profits via higher prices." Goldstein did not add that such tactics cut into farmers' profits by raising fertilizer prices.

This extraordinary market concentration has essentially given the fertilizer barons leverage to influence the prices they can charge farmers. The price of corn and the price of fertilizer fluctuated more or less in tandem until about 2003. That's when the industry went through a wave of mergers. Fertilizer prices began to shoot up faster than corn prices during the 2003–12 boom, then fell much more gently during the bust that followed. This divergence demonstrates a concentrated industry's ability to push big price hikes when its main customers are flush and keep those high prices "sticky" when its customers hit a rough patch.

Then there are seeds and pesticides. Until the mid-1990s, seeds and pesticides were wholly separate industries. But in the 1980s, a massive chemical company with a robust pesticide division, Monsanto, jumped into the emerging field of crop biotechnology. That is, it began investing profits from selling chemicals into the then budding field of using gene transfer to create novel crops.

Monsanto executives quickly realized that the surest path to big profits was to engineer widely planted crops to resist its own blockbuster pesticide Roundup, which was then under patent. At first, the idea was to monetize its crop-biotech triumphs by licensing patented traits to seed companies, as though traits were software and seeds were hardware. In the end, they decided to capture maximum profits from the genetically modified organism (GMO) enterprise by jumping into the seed business.

In 1996, Monsanto dropped $240 million to buy Asgrow, a major soybean seed supplier to Midwest farmers, and another $319 million for a 40 percent stake in corn-seed supplier Dekalb. The floodgates opened. By 1999, rival DuPont—which had also invested heavily in GMOs—had subsumed the biggest Midwest seed supplier of all, Pioneer Hi-Bred International, for $7.7 billion. And so began a spasm of mergers that, by the mid-2010s, had fused seeds and chemicals into a single industry, led by a "Big Six" of agrichemical-seed companies that towered over farmers in the Midwest and globally alike.

And then in 2015, with crop prices low and shareholders demanding bigger profits, another spate of earth-shaking mergers began, and the Big Six firms collapsed into the Even Bigger Three. Now, just three massive firms—Corteva Agriscience (the ag spin-off of the merger of chemical behemoths Dow and DuPont), Chinese-owned Syngenta, and Bayer, which in 2018 completed its takeover of Monsanto—control nearly 60 percent of the entire $68 billion global seed market, which includes both conventionally bred and GM seeds. While they merely dominate the non-GMO seed market, their grip over the GM seed trade is near total.

In the U.S. Corn Belt, the Big Three loom even larger, controlling around 80 percent and 75 percent of the U.S. corn and soybean seed markets, respectively. Their patented genetic traits appear in close to 100 percent of these two crops.

To bring forth the massive annual corn and soybean harvest that essentially operates as the raw material for our food system, farmers rely on herbicides, insecticides, and fungicides—produced almost entirely by the same three companies.

Corteva, Syngenta, and Bayer also control 60 percent of the $2.9 billion global pesticide trade. Having engineered corn and soybeans to resist their herbicides, these corporations also aggressively market seeds coated with their own and each other's insecticides and fungicides, which then suffuse the crops, killing bugs and mold. As Pat Mooney, a longtime industry watcher and founder of ETC Group (Action Group on Erosion, Technology and Concentration), put it in a 2019 report, "7,000 plant breeding entities have effectively become four; 65 pesticide producers have become nine; [and] seeds have merged with pesticides."

As with big-ticket farm equipment, the hyper-consolidation of seed suppliers and the companies' heavy marketing of expensive GM traits have put upward pressure on prices. This forces farmers to spend more money on seeds as a percentage of the profits they receive for their harvest. In the mid-1990s, just before the introduction of GMOs, corn farmers' seed costs hovered at around 8 percent of their revenue. By 2004, when GM corn seed had largely overtaken the market, seed costs had reached 16 percent of revenue—a level that has largely held ever since. The seed/pesticide industry has created a closed-loop system: proprietary seeds engineered to resist herbicides made by the industry, coated with insecticides and fungicides also made by the industry.

The rise of what is known as "digital" or "precision" agriculture threatens to tighten the loop still further. On its face, precision ag is a benign idea. Using information technology—sensors attached to combines, satellite- and drone-captured imagery, advanced weather forecasts—farmers can generate loads of data that can help shape planting and spraying decisions, potentially reducing the use of pesticides and fertilizers. In a 2019 study, Michigan State University gauged crop yields across the Corn Belt over ten years using these very tools: satellite imagery (which can predict crop yields by the quality of light reflected from fields in high summer) and harvest data from sensors attached to combines at a patchwork of farms across the region. They found that fully 26 percent of the region's land—31 percent in Iowa—delivered consistently low crop yields. Yields were so low in these areas, the study's lead author, Michigan State environmental sciences professor Bruno Basso, told me, that farmers were spending more money on fertilizers, seeds, pesticides, and other farming expenses than they made from selling their crops. Farmers would be better off, Basso said, simply fallowing this land than farming it. They could also simply scale back their chemical applications on these low-production acres.

Both fallowing and reducing chemicals on these areas, however, would cut into the seed-pesticide companies' bottom lines, which is why in recent years these firms have spent mightily to control access to precision ag information sources. In 2013, Monsanto spent $930 million to snap up a Silicon Valley weather start-up called the Climate Corporation, which uses publicly funded National Weather Service data. That move started an avalanche of investment by the seed-pesticide industry in this new technology, and by 2016, Monsanto, Dow, DuPont, and Syngenta had all rolled out their own precision ag arms. Hand your field data over to us, went their pitch to farmers, and let us crunch it and combine it with our data drawn from satellites and drones. Then we'll tell you what seeds to plant, how densely to plant them, and which of an array of weed killers, bug killers, and fungus killers to spray on them.

Monsanto's Climate Fieldview program quickly took the lead. By 2016, the company was claiming that farmers representing 45 percent of U.S. corn and soybean production had adopted the service, most of them opting for the basic (free) version. A year later, even though Climate Fieldview was generating no profits, then CEO Hugh Grant cited it as the company's crown jewel in a presentation to investors, placing it at the heart of an "integrated solution strategy that brings seeds, traits, chemistry and data science tools to farmers around the world."

Robert Fraley, Monsanto's longtime chief technology officer, expanded on these themes in a 2016 interview with the trade magazine *AgFunder News*. Asked to envision the company in ten years, he put Climate Fieldview at the center of a vision of Monsanto as a kind of Amazon for farmers, seamlessly providing everything they need. He argued in essence that the agriculture industry isn't consolidated *enough*; bewildered farmers were too often forced to purchase their seeds, pesticides, fertilizers, and equipment from different entities, he said. He painted a future in which farmers would essentially outsource their decisions to Monsanto—or at least rely on the company to narrow their choices dramatically. With Climate Fieldview as the hub, Monsanto would evolve into a company that could "integrate it all" to "optimize all of those discrete decisions into an ideal system to maximize crop production and profitability." Merging the surveillance apparatus of Big Data with Monsanto's seed catalog and breeding wizardry, the company would be able to tailor seed genetics directly to meet a farmer's field conditions, and then coat the seeds with "multiple products to help control bugs and weeds" as well as soil-dwelling microbes to "stimulate the growth of that plant."

The same year Fraley gave that interview, Monsanto shareholders agreed to sell the company to Bayer for $66 billion. But the Climate Fieldview dream burns bright. Farmers with combines made by Deere and AGCO—which together control 70 percent of the U.S. combine industry—can opt to supply harvest field data through sensors straight to Bayer, which in turn can "export prescriptions" right back to the combine, advising what to spray and how much of it. By 2019, Fieldview products were being used on nearly 60 million acres of U.S. corn and soybeans—about a third of the land devoted to the crops.

Bayer's two main rivals have moved into the business, too. Through its digital ag subsidiary Granular, Corteva Agriscience has a deal with Planet—"an integrated aerospace and data analytics company"—that "will integrate Planet's industry-leading daily, global satellite imagery data into Granular's farm-management software suite." To bring that data down to earth, and into farmer's combines, Granular also has a deal with John Deere. Syngenta, too, has a digital ag arm which customers can opt to sync with John Deere equipment. Syngenta offers farms AgriEdge Excelsior, described as a "whole-farm management system service," complete with a satellite offering and a John Deere combine deal.

As Basso's study suggests, rationalizing corn belt agriculture with satellite data and on-farm sensors *could* actually empower farmers to make better decisions. But the farmers' interests and the industry's don't necessarily align. The companies offering "whole-farm management system service" and "integrated solution" strategies are also in the business of selling patented seeds and brand-name pesticides—not unlike a financial adviser peddling her own company's financial products to consumers, whose interests she is nominally meant to protect. Then there's the fact that to make digital agriculture hum, the highly concentrated seed-pesticide industry has to collaborate with the dominant equipment company, Deere. In this way, two of the main oligopolies farmers deal with cooperate to gather vast amounts of on-farm data. As a Facebook user might have asked in 2010—before we understood the value and potential for abuse involved in handing loads of personal information across populations to a single company—what could possibly go wrong?

Once they've harvested their crops, farmers need to find markets for them. Here, they hit up against yet another oligopoly: the grain traders, globe-spanning firms that buy commodities from farmers and ship them to buyers, whether domestically or on every continent on the globe. The global grain-trading market has for decades been tightly controlled by a handful of companies known as the ABCDs, for U.S.-owned giants Archer Daniels Midland, Bunge, and Cargill (the largest privately held company in the United States) as well as the Netherlands-based Louis Dreyfus Company.

U.S. corn farmers ultimately choose among three megabuyers: ADM, Cargill, and Ingredion (formerly Corn Products International). These companies control 87 percent of the market, according to a 2014 study by the University of Missouri rural sociologist Mary Hendrickson. For the U.S. soybean crop, ADM and Bunge are the top two buyers, Hendrickson found. Together with Cargill and Nebraska-based Ag Processing, they process 85 percent of U.S.-grown soybeans.

Grain markets are global—U.S. corn and soybean farmers compete with their counterparts across the world, most prominently in Brazil, where soybean production has jumped to near U.S. levels in recent years and corn output has surged. Whereas ADM is the top player in U.S. grain markets, Bunge is king in Brazil, especially in soybean processing. Bunge processes 27 percent of the Brazilian soy crop, while ADM owns a relatively small 10 percent. That farmers in Brazil and the Corn Belt sell their product to the same handful of buyers ramps up their competition with one another, putting downward pressure on prices.

The grain traders don't just buy corn and soybeans from farmers and sell it to end users like meat producers. They also transform it into a variety of other highly profitable products. ADM and Cargill are both heavyweights in the U.S. food-ingredients industry, for example.

ADM is also the number two U.S. producer of corn-based ethanol. Its lobbying efforts, starting in the late 1970s, paved the way for the government mandates decreeing that around 15 billion gallons of ethanol be blended into the nation's fuel supply, offsetting around 10 percent of U.S. gasoline use. Stunningly, this use alone consumes a third of the entire U.S. corn crop. Industry proponents and the U.S. Department of Agriculture insist that turning corn into car fuel lowers fossil fuel use and cuts back on carbon emissions, but these claims are fiercely contested.

Cargill, for its part, has used its status as a crop megabuyer to push its way into yet another cartel that looms over Corn Belt farmers: meatpacking. The firms in this business buy animals from farmers, slaughter them, and then cut them into products for the supermarket counter and for institutional buyers like restaurants, schools, and hospitals. Since the great bulk of U.S.-grown meat comes from animals fed with corn and soybeans, meatpacking is yet another industry that lives off the Corn Belt's bounty—and it may be the most oligopolistic of all.

"Feeding animals and selling meat have become the biggest sources of profit for Cargill," the *Financial Times* reported in 2018, accounting for the bulk of the firm's $3.1 billion annual profit that year. Cargill alone slaughters and packs nearly one in four beef cows raised in the United States. Its few rivals in the U.S. beef-packing market are made up of the globe's biggest meat companies. The Brazilian giant JBS, which burst onto the U.S. meat scene with a series of acquisitions starting in 2007, also owns about a quarter of the U.S. beef market.

Arkansas-based Tyson enjoys roughly the same U.S. beef market share as Cargill and JBS. Finally, National Beef—which is jointly owned by the investment bank Jefferies Financial Group and yet another Brazilian meat titan, Marfrig—owns about 10 percent of the market. Together, Cargill, JBS, Tyson, and National Beef account for 85 percent of the U.S. beef market.

Tyson and JBS are also major players in the U.S. pork market, making up two of the four firms that slaughter nearly 70 percent of U.S. hogs. Their share of this market is topped only by that of Virginia-based Smithfield, the globe's biggest pork packer, a subsidiary of Chinese meat conglomerate WH Group. Tyson and JBS also occupy the top two slots in chicken processing. Tyson alone slaughters and packs 20 percent of the chickens grown in the United States, while JBS's market share clocks in at around 15 percent. These two companies, along with Sanderson Farms and Perdue, crank out nearly half of U.S.-grown chicken.

The international nature of the oligarchy that dominates the U.S. meat business—with its ownership groups in São Paulo (JBS, Marfrig) and Beijing (Smithfield)—is no accident. In short, the United States has emerged as a prime spot to raise and process cheap meat to be consumed elsewhere. It's not hard to see why. U.S. meat consumption exploded in the post–World War II years—a pattern that can be directly tied to ever-bigger corn and soybean harvests and thus cheaper feed—but then leveled off at a very high rate around 2000. Today, we still eat massive amounts of meat: about 220 pounds annually per capita, or more than a half pound per day. That's more meat per capita than all but a handful of smaller countries. But our appetite for meat hasn't budged in twenty years; and slow U.S. population growth, combined with flatlining per-capita demand, makes for a large but stagnant market.

With their business models geared to steady growth in the demand for goods, U.S. meat companies could no longer count on the domestic market alone. So they began looking outward for growth, particularly to China and the rest of Asia, where meat consumption is on the rise. In a 2008 interview with journalist Christopher Leonard, then Tyson CEO Don Tyson lamented, "We have done about as much in the United States as we can do." Leonard summarized Tyson's emerging strategy like this: "Now the company plans to duplicate that [past] success in developing nations where a growing middle-class population for the first time can afford to eat meat and visit drive-through windows."

Between 2009 and 2018, U.S. beef exports jumped by 50 percent and pork exports by 30 percent. Big Chicken's ascent came earlier. In 1990, U.S. producers were exporting around 1.1 billion pounds of chicken per year. By 2007, exports had reached 7.2 billion pounds, a level that has held fairly steady since. Overall U.S. meat exports surged from less than 10 billion pounds in 2000 to more than 16 billion pounds in 2018. As of 2018, more than 25 percent of U.S. pork, nearly 17 percent of U.S. chicken, and 12 percent of U.S. beef is sent to foreign markets.

Normally, filthy industries like factory-scale farming have a way of drifting from the United States to countries further down the GDP ladder, where wages are lower and environmental and labor regulations aren't as rigorous. Why has factory animal farming maintained a vast and growing footprint here? The short answer is that corn and soybeans are so abundant and cheap that the United States maintains a comparative advantage. In a 2012 study, U.S. Department of Agriculture researchers found that U.S. operations churn out hogs for about $0.57 per pound, versus $0.68 per pound in China's own growing set of factory-scale hog farms. The main difference is feed costs. U.S. pig producers spend about 25 percent less on feed than their Chinese counterparts do, the USDA found, because the "United States has more abundant land, water, and grain resources." As for the U.S. environmental regime and labor laws, they aren't strong enough to hinder the industry's advance. In essence, the Midwest is emerging as the industrial-scale farm for China and other countries with rising demand for meat.

So even as U.S. consumers hold steady in their appetite for meat, domestic production is expanding. As of 2019, the poultry industry was in the midst of a ramp-up, with seven new large-scale slaughter plants under construction, including facilities for industry giants Tyson and Sanderson. Retail colossus Costco is also barreling into the chicken game, breaking ground on a $300 million plant in Nebraska that will produce almost 100 million birds a year. Of course, those will be transformed into "rotisserie chickens" to be enjoyed by domestic Costco customers, but it will also free up more chicken for export. In 2018, a Sanderson executive told *Bloomberg News* that the chicken industry's production will likely expand by 3 percent annually from 2019 through 2021, "the most since a three-year expansion ended in 2005." The pork industry is growing, too, with a massive new packing plant debuting in Iowa in 2019 and another one rolling out in 2020.

All this expansion, and the associated environmental wreckage of industrial-scale meat production, is premised on cheap feed from the Corn Belt's farms. So the region's annual harvests underwrite a growing global meat-production empire here in the United States—ensuring more demand for the Corn Belt's crops, even as U.S. demand for meat levels off.

In a bit of economic turnabout, China is in essence taking advantage of lax environmental and labor standards and cheap raw materials in the United States to supply its citizens with something they can't get enough of. The Corn Belt is to industrial pork as Shenzen is to the iPhone.

It's difficult to see what the Corn Belt gains from this trade, beyond profits for the meat giants (many of which are foreign-owned). It certainly doesn't deliver broad-based prosperity to the regions where the animals are grown and slaughtered. In their excellent recent books, the journalists Christopher Leonard (*The Meat Racket: The Secret Takeover of America's Food Business*) and Ted Genoways (*The Chain: Farm, Factory, and the Fate of Our Food*) have demonstrated how this quasi-cartel of meat giants uses its market heft to squeeze the farmers who raise those billions of steers, hogs, and pullets, as well as the workers who take on the unenviable task of slaughtering them by the millions.

Once a largely unionized middle-class profession, slaughterhouse work has emerged as low-paid and startlingly dangerous work. Drawing an average hourly wage of $13.38, meatpacking workers incur injury and illness at 2.5 times the national average. The prevalence of repetitive-motion conditions among laborers is nearly seven times that of other private industries. Much has to do with the speed at which they work: hog carcasses weighing as much as 270 pounds come at workers at an average rate of 977 per hour, or about 16 per minute. It's no wonder that refugees fleeing political violence have emerged as an important labor source for the industry. (On the other hand, foreign-born meatpacking workers, who make up at least half the industry's labor force, have breathed new life into Corn Belt towns that have been hollowed out by industrial agriculture.)

Nor does the meat-production machine even use up the Corn Belt's vast annual crop output. Nearly half of U.S. soybeans and 15 percent of corn are exported, almost entirely to supply U.S.-style concentrated livestock operations abroad. According to agrichemical industry and U.S. government dogma, this outflow of feed crops provides a crucial humanitarian service: "feeding the world."

But as the Environmental Working Group showed in a 2016 report, the twenty countries that take in 86 percent of U.S. food exports are almost all highly developed (based on U.N. measures of life expectancy, income, and level of education) and have low hunger rates. Meanwhile, the nineteen nations with the biggest and most dire hunger problems import very little food from the United States—they accounted for a whopping 0.5 percent of total U.S. agricultural exports in 2015, the Environmental Working Group reports. Even accounting for food aid, the great U.S. ag behemoth contributes very little to feeding the poorest of the world's nations, providing on average just over 2 percent of their food supply.

Indeed, as the environmental scientist Jonathan Foley has shown, devoting huge swaths of land to feed crops like corn and soybeans is monstrously inefficient. He writes:

> For corn-fed animals, the efficiency of converting grain to meat and dairy calories ranges from roughly 3 percent to 40 percent, depending on the animal production system in question. What this all means is that little of the corn crop actually ends up feeding American people. It's just math. The average Iowa cornfield has the potential to deliver more than 15 million calories per acre each year (enough to sustain 14 people per acre, with a 3,000 calorie-per-day diet, if we ate all of the corn ourselves), but with the current allocation of corn to ethanol and animal production, we end up with an estimated 3 million calories of food per acre per year, mainly as dairy and meat products, enough to sustain only 3 people per acre. This is lower than the average delivery of food calories from farms in Bangladesh, Egypt and Vietnam.

As for the cuisine that grows from transforming corn and soybeans into cheap meat, sweeteners, fats, and processing agents, it's hardly an advertisement for robust health and longevity. Nearly 60 percent of the calories Americans consume come from the very "ultra-processed" foods that are shot through with corn and soybean derivatives. On any given day, about

37 percent of Americans have at least one meal from a fast-food chain—another industry puffed up by the Corn Belt's bounty (think corn- and soybean-fed beef, corn-sweetened soda, potatoes fried in soybean oil). It's no wonder that "nearly half of all American adults have one or more chronic diseases that are related to poor quality diets," a 2017 study by National Cancer Institute researchers reported. The so-called western diet—the culinary manifestation of the corn/soybean duopoly—is a harbinger of such diet-related distress, and researchers have documented that as it spreads across the globe, it leaves a pandemic of diabetes, obesity, and heart disease in its wake.

In short, our country's gargantuan corn and soybean crops, concentrated in Iowa and surrounding states and occupying more than half of U.S. farmland, is essentially a zero-profit industry for farmers, propped up by billions in government payouts. The main beneficiaries are a set of interlocking, enormous corporations, each generating billions of dollars for shareholders and delivering in exchange a mountain of health-ruining food.

In 2018, President Donald Trump launched a multifront trade war against several important buyers of U.S. agricultural commodities. The move antagonized China, which retaliated with stiff tariffs on pork and soybeans; and Mexico, which did the same with corn. Eschewing U.S. farm goods, China's buyers ramped up purchases from emerging corn and soybean powerhouses like Brazil and Argentina to meet their needs, and Corn Belt farmers saw already-low prices drop, pushing them deeper into crisis. Farm bankruptcies in the region jumped by 19 percent in 2018, reaching their highest level in a decade and nearly twice the 2008 rate. In 2019, as the trade battle lurched on, chaos and uncertainty in these once-booming markets put yet more downward pressure on corn and soybeans. Farm debt reached levels last seen during the brutal 1980s farm crisis.

But input suppliers like Nutrien, Syngenta, and Bayer, and grain buyers like Archer Daniels Midland and Bunge, are just as dominant in

South America as they are in Iowa, so Trump's trade policies caused them minimal disturbance. The meat industry is similar—companies like Smithfield and JBS can easily cash in by sending China pork and chicken grown at its outposts in Poland and Brazil rather than in Iowa and North Carolina. In such globalized markets, farmers lose, and the meat and grain giants win.

It's worth digging into what the global span of the Big Ag oligopolies means for Corn Belt farmers, all locked in a hypercompetitive market, pitted against hundreds of thousands of their peers, who are clumped together around them and producing the same commodities. Everyone strives to grow as much as possible. When they succeed en masse, prices drop; everyone loses, except for input sellers and crop buyers. Smaller, more diversified farms go broke and sell out; bigger farms that laser-focus on corn and soybeans can leverage economies of scale and eke out a profit, buoyed by crop and insurance subsidies. It's no surprise that the region's farms exist in a state of steady attrition, driven forward by the hope of another boom just around the corner.

The 2017 U.S. Census of Agriculture tells the story. As recently as 1987, more than 25,000 Iowa farms—about a quarter of the total, representing more than 500,000 acres—included oats in their crop rotations. Breaking up the corn-soy duopoly with a "small grain" like oats delivers all manner of ecological benefits—slashing fertilizer and pesticide needs. By 2017, just 15 percent of Iowa farms grew oats, and oat acres had plunged to 36,580—an astonishing 92 percent drop. Wheat, another ecologically beneficial small grain in the Corn Belt context, showed a similar trend. Over the same period, corn and soybean acres jumped by more than 20 percent each, reaching a combined 21 million acres.

Diverse crop rotations require more skill and attention, and are best suited to midsize farms that ideally provide farmers with enough scale to make a profit without being so big that they need to be streamlined to manage. Farms that fit that description—between 180 acres and 999 acres in size—experienced an extended depression over the past three decades, falling from 45,000 in 1987 to about 29,500 in 2017, a 35 percent drop. During the same period, farms over 2,000 acres in size jumped from 320 to 1,890—a 490 percent increase.

Iowa saw its midsize, diversified farms wither, while mega-farms specializing in corn and soybeans proliferated. Similar trends prevailed in other Corn Belt states.

As farm numbers shrank, surviving farmers necessarily had fewer peers competing for those few corn and soybean buyers. You might think the attrition would win the survivors some hard-earned leverage against the oligopolies that surround them. Instead, as the U.S. Corn Belt experienced a slow exodus, industrial-scale corn and especially soybean farming took root and expanded rapidly in Brazil and Argentina. And a close look at big agriculture in South America—particularly in Brazil's Cerrado region—makes abundantly clear who drives the corn and soybean industries, and where their priorities ultimately lie.

In Brazil, much attention has been focused on the expansion of cattle ranching and soybean production in the Amazon rainforest. In the mid-2000s, the grain-trading and meat giants signed voluntary agreements not to buy goods from Amazon land deforested after 2006 and 2009 for beef and soybeans, respectively. These much-lauded moves indeed slowed the pace of Amazon deforestation—but sped it up in another vast region with robust biodiversity.

When Portuguese settlers arrived in Brazil in the fifteenth century, they immediately began forcing enslaved Africans and indigenous people to clear-cut the coastal rainforest for sugar and coffee production. But a vast swath of the Brazilian interior resisted colonial agriculture: a scrubby savanna region, composing about a fifth of Brazil's territory, that the settlers dubbed the Cerrado—Portuguese for "closed." For the purposes of Portugal's rulers, the Cerrado was a massive wasteland; its acidic soil couldn't support the desired crops. As with most so-called wastelands, the Cerrado had been quite useful for a wide variety of beings for millennia. The region had already supported human populations for at least twelve thousand years. Until abolition in 1888, fugitive enslaved people fled into the Cerrado and created settlements called *quilombos*. Even now, thirty years after the advent of industrial agriculture, dozens of indigenous groups live there, as do more than three thousand *quilombola* communities.

A stunning variety of flora and fauna evolved to thrive in the warm, intermittently wet climate. "The Cerrado hosts almost 12,000 species of native plants, about 251 species of mammals, 267 species of reptiles, 209 species of amphibians, and a rich and diverse population of about 850 bird species, all distributed over a wide variety of habitats," according to researchers from Brazil's Center for Earth System Science. The World Wildlife Fund calls it "one of the richest places of biodiversity in the world," and it is widely considered the globe's most biodiverse savanna.

While the shrubs that dominate the Cerrado's landscapes don't trap quite as much carbon as the Amazon's grand trees, they make up a kind of "inverted forest" by plunging deep roots into the acidic soil in search of nutrients and moisture. So they store three quarters of their biomass—and thus loads of carbon—underground. And the Cerrado plays a vital role in the health of other ecosystems: numerous important tributaries of the Amazon River originate in the Cerrado, and its waterways feed the world's largest wetland, the Pantanal, as the World Wildlife Fund researchers outline.

In the early 1980s, scientists discovered that treating the Cerrado's soil with lime, along with heavy doses of mined phosphorus fertilizer, made it hospitable to U.S.-style industrial farming. What followed was an historic agricultural boom, on a grander scale than any other on Earth since the plowing of the Midwest prairie in the nineteenth century. Its financiers and large landholders include the Brazilian soybean magnate and politician Blairo Maggi, über-investor George Soros (through farmland fund Adecoagro, whose other owners include the Qatar Investment Authority and the Dutch pension fund PGGM), and wealthy European heir Jacob (Lord) Rothschild. Massive U.S. investment funds have also taken a bite, including those familiar big California landholders TIAA and the Harvard endowment. Their Cerrado holdings have embroiled both funds in charges of benefiting from land grabs.

In its entirety, the Cerrado covers 494 million acres—an area that's equivalent to almost five Californias and about 20 percent larger than all the land under crop production in the United States. Already, more than 60 percent of that land has succumbed to agriculture.

In a 1997 paper, Norman Borlaug—a Nobel laureate and the father of the so-called green revolution that spread U.S.-style farming to India and across the world—extolled the plowing of the Cerrado, then in its infancy. In a single sentence, he crystalized a utilitarian view of landscapes and a messianic belief in the gospel of better living through chemistry:

> It should be emphasized that the vast Cerrado ecosystem, over untold millions of years, was leached of nutrients and made unproductive by Mother Nature, whereas over the last four decades, it is being converted into a huge new, highly productive "bread basket" through the collective efforts of scientific man, effective government and private sector programs, plus aggressive, creative farmers and ranchers.

The climate implications of the transformation are huge. In a 2017 paper, a group of U.S. and Brazilian researchers calculated that between 2000 and 2013, carbon emissions from expanding cropland in the region averaged 16.3 million metric tons per year—equal to a year's tailpipe carbon emissions from more than 12 million cars, roughly twice the size of greater Los Angeles' car fleet. Of course, deforestation for agriculture lands a one-two punch to the climate: plowing vaporizes carbon stored over millennia and nixes future sequestration, because annual crops grab much less CO_2 than native perennials. (A similar transfer of carbon from the soil to the atmosphere occurred during the settling of the U.S. prairie in the nineteenth century.)

And these Brazilian mega-farms, with their wealthy owners and recently broken land, are the direct competitors to farmers in the Corn Belt, vying with them in a race to the bottom to sell corn and soybeans to China and other fast-growing countries.

Cerrado farms capitalized when, in 2005, U.S. corn prices began an upward swing, powered by an ethanol-boostering federal energy policy that pushed up the value of corn on global markets. In the northern Corn Belt climate, farmers can plant only one cash crop per year; but in the tropics, farmers can squeeze in a second. As prices rose in the middle 2000s, Cerrado farmers began to plant corn after soybeans. Between 2005 and 2015, Cerrado corn production doubled. After the "extraordinary increase in Brazil's corn production over the past decade," the USDA reported in 2015, Brazil "emerged as the largest U.S. competitor in the global corn market with second-crop corn, harvested late in the local marketing year, boosting exports from September to January, months traditionally dominated by Northern Hemisphere exporters."

As for soybeans, Brazil overtook the United States as the supplier of choice to China, the destination for two thirds of the world's soybean exports, a half decade before Trump became president and started launching trade wars. In 2012, U.S. farmers supplied 45 percent of China's soybean market, and their counterparts in Brazil, 40 percent. Every year since, though, Brazil has held the lead. In 2017, before Trump opened trade hostilities, Brazil had 53 percent of the market, versus the United States' 34 percent.

But Cerrado farmers confront the exact same handful of seed-chemical, fertilizer, meat, and grain-trading companies as their Corn Belt peers do. For a Bayer exec looking to boost seed or herbicide sales, or a Cargill trader hoping to profit from moving soybeans to China, it makes little difference if the beans are planted in the Cerrado state of Mato Grasso or in Iowa, or if they're shipped from New Orleans or Santos, Brazil.

For global agribusiness, the Cerrado has no value as homeland, habitat, carbon sink, or source of clean water. What it represents is a vast store of land to be put to use to expand production of the crops it favors: commodities that require heavy doses of agrichemicals and that can't be consumed directly but must first either be highly processed, used to fatten animals, or consumed by cars instead of people in the form of biofuel. In short, boosting sales of chemicals while keeping crop prices down. And the existence of the Cerrado—and also of newly emerging corn-and-soybean bastions like Ukraine's "black soils" heartland—mean fierce and ever-accelerating competition for Corn Belt farmers.

For the companies' executives and shareholders, the arrangement is excellent. Monsanto's last CEO before being subsumed by Bayer in 2018, Hugh Grant, tells the story. Between 2000 and 2018, the company's share price rose by a factor of fourteen. By 2017, his last full year, Grant was bringing home total annual compensation of $19.7 million (a third higher than the average Fortune 500 CEO's salary); and on his retirement he departed with a golden parachute worth $77 million.

For U.S. consumers and farmers, the value proposition is murkier. Consumers encounter bountiful but health-ruining calories from cheap meat, sweetener, and fat. Farmers attain the means to produce bumper crops and access to global markets, even as robust, stable profits ultimately elude them, flattened out by fierce competition with peers in other regions. It's the land—a precious store of essentially irreplaceable topsoil—that unambiguously loses. Soil formed over millennia in the former prairies and wetlands of the Midwest isn't an ephemeral commodity that can be replaced the following year, like a bushel of corn or a hog raised in a CAFO. It's one of the jewels of global agriculture, and we're drawing it down, even as climate change accelerates the process and makes this essentially fossil resource ever more precious.

5

Failing Upward

In 2016, I entered what was arguably the global seed and agrichemical industry's greatest citadel: Monsanto's global R&D headquarters, a sprawling complex located amid the corn fields at the edge of St. Louis.

It was a year before Monsanto shareholders agreed to sell the company to German chemical giant Bayer and two years before the deal fully closed. At that point, Monsanto was a fixture in the financial press for its ruthless, ultimately failed quest to take over rival seed-pesticide giant Syngenta.

Under the natural sunlight and amid the potted plants of the visitors' atrium, with its soaring windows and exposed-brick walls, I was soon clasping hands and exchanging grins with Robert Fraley, a giant in the world of agricultural biotechnology. As a young microbiologist in the early 1980s, Fraley led the research team for Monsanto—then a company much more focused on making industrial chemicals than farm inputs—that created one of the world's first genetically modified plants. By the time I met him, Fraley was in his mid-sixties and a full-on corporate titan, having spent years as the chief technology officer of a globe-spanning company with $15 billion in annual revenues. Bald, bespectacled, sturdy, he was dressed in modern corporate-casual: khaki trousers with a blue quarter-zip fleece.

He regarded me with an avuncular air: a decorated scientist prepared to indulge the ignorance of someone who simply needed to be brought round to the truth with reason. He told me he was eager to engage with a public critic of his company in particular and ag biotechnology in general. I told him I wanted to hear his side of things.

Honestly, I was a bit starstruck. For years, first in a personal blog, later in *Grist Magazine*, and then in *Mother Jones*, I had been a fierce industry critic, documenting in dozens of columns and stories its bigfoot tactics with farmers, its relentless marketing of pesticides, its specious presentation of itself as the savior of the world's hungry, and its ever-tightening grip over a vital common resource, seeds.

I wrote often about Monsanto because it was the most aggressive and pervasive of the six transnational behemoths that then dominated the global trade in seed and pesticides. In the mid-1990s, the company rolled out blockbuster genetically altered seeds that transformed agricultural biotechnology from a speculative venture into a $17 billion global market.

For most of my previous years as a food politics writer, my relationship with the company had been contentious. Indeed, my first break in the field came in 2005, when a Monsanto attorney emailed the small farm where I was then working demanding that I stop using the name of its blockbuster product, Roundup Ready, in a sarcastic way on my blog. I published the cease-and-desist letter and a defiant response; the post went viral. (Monsanto never responded to my riposte.)

So why, a decade later, had the company invited a pesky critic into its inner sanctum?

No company has loomed over and shaped Corn Belt agriculture like Monsanto, the first to commercialize genetically engineered crops. Its signature innovation—crops tweaked to withstand Monsanto's own weed killer, Roundup—transformed the seed and pesticide industries in its own image, forcing rivals like Dow and DuPont (whose agriculture divisions are now combined into a single entity called Corteva) to imitate its business model and pay up to license its traits. And the rise of Roundup Ready corn and soybeans offered farmers unprecedented ability to devote huge areas of land to monocrops without having to worry about weeds, even as Monsanto and its few peers claimed an ever-higher share of their revenues. The technology also dramatically escalated the chemical war farmers had been waging against their biological pests since the post–World War II dawn of industrial agriculture.

To understand the dilemmas faced by Corn Belt farmers as they navigate a changing climate, it's crucial to understand how Monsanto arrived as such a dominant position in our food system, which it maintains as a subsidiary of Bayer.

Before it emerged as an agribusiness giant in the 2000s, Monsanto was a chemical powerhouse—one that rode the mid-century better-living-through-chemistry boom with a product line that included both pesticides and industrial chemicals. The company established a record for developing blockbuster products and defending them vigorously when they proved to harm people. In 1976, Congress banned the highly carcinogenic industrial coolant PCB, over which Monsanto had enjoyed a highly profitable monopoly in the United States for almost half a century. Documents uncovered in a lawsuit showed that Monsanto had known about the dangers of PCB exposure for years before the ban. The company ultimately agreed to pay a $390 million settlement to victims. Monsanto was also known for marketing the artificial sweetener aspartame and the artificial grass known as AstroTurf, both of which generated significant public outcry. Then, in the early 1960s, still riding high on the industrial chemicals wave, Monsanto rolled out a division to develop products for a subset of a fast-growing market: agriculture.

The herbicide market barely existed before World War II. At the time, Corn Belt farmers relied mainly on cultivation to kill weeds. Crop rotation was also often used to disrupt weed patterns. During the war, the U.S. and U.K. governments poured resources into developing defoliants as a weapon, the idea being to weaken an enemy nation by wiping out its crops. While the effort didn't succeed in time to deliver weaponry for that particular war, it did yield a plant-killing compound called 2,4-D that was quickly marketed by rival chemical companies, including Dow, to farmers—and taken up in the Corn Belt; 2,4-D was particularly good at killing the broadleaf weeds that bedeviled Iowa farmers at the time. The advent of 2,4-D meant two things in the region. One was that crop rotation as a strategy to control weeds and other crop pests became obsolete, allowing farmers to move away from crops like oats and wheat and focus on corn and soybeans. The other was that while 2,4-D knocked out broadleaf species, it left grassy ones mostly intact. And so with their broadleaf peers wiped out, grassy weeds proliferated—creating a need for new chemicals that could effectively kill them.

The market for herbicides exploded. In 1940, U.S. farmers spent less than $2 million total on weed-killing chemicals. By 1962—when Monsanto initiated its agricultural unit—herbicide expenditures exceeded $270 million. That same year, Rachel Carson documented the ecological downsides of pesticides for the landmark book *Silent Spring*. Her book had a massive cultural impact and turned the tide against DDT (another Monsanto moneymaker), but did little to hinder the use of most compounds designed to poison the insect, fungal, and weed pests that bedevil farmers. Between 1960 and 1970, sales of U.S. farm pesticides (a category that encompasses insecticides, herbicides, and fungicides) spiked by a factor of three in inflation-adjusted terms, and they rose steadily, though less dramatically, in the decades that followed.

And 2,4-D didn't triumph just in farm fields. Conceived as a chemical weapon in the 1940s, it emerged as a defoliant deployed by the U.S. military in Vietnam in the 1960s—both on its own and combined with another, even more toxic herbicide in the form of Agent Orange. Produced by Monsanto, Dow, and a few other chemical companies, Agent Orange inflicted massive ecological and human damage in Vietnam, among U.S. soldiers and Vietnamese civilians alike.

Monsanto's agriculture division didn't score its first exclusive blockbuster farm product until the early 1970s, when company researchers noticed that a water-softening chemical they were working on had a tendency to kill plants. They tossed it to their colleagues on the agriculture team, and there, a young chemist named John Ganz tweaked it into a weed-slaying dynamo that came to be called glyphosate. Herbicides then on the market tended to have what agronomists call a "narrow spectrum": they killed only some types of plants. (Think of 2,4-D attacking broadleaf species but not others.) They also tended to be toxic to most forms of life—hazardous not just to weeds but also to fauna that crossed their path.

Glyphosate, it seemed, was the opposite. According to Monsanto, it combined broad-spectrum destruction—the ability to kill almost any plant, broadleaf and grassy alike—with low toxicity to other life forms. It works by jamming up plants' ability to produce an enzyme necessary for making vital amino acids, the building blocks of protein. That's a strategy for nourishment that plants share with fungi and bacteria but not insects, birds, fish, or mammals, all of which simply consume protein. Since the chemical's weed-flattening "mode of action," to use the agronomists' term, doesn't apply to people, it must be nontoxic. So the thinking went.

Monsanto first rolled out its glyphosate-based herbicide in 1974, naming it Roundup to stress its effects on a field's entire population of weed species. It caught on rapidly. By the mid-1980s, it had emerged as the most profitable product in Monsanto's agriculture division.

As Roundup gained traction, the field of biotechnology—manipulating genomes to create novel products and organisms—was undergoing a massive investment boom. Successful gene-splicing experiments in the early 1970s had piqued interest in venture capital circles. The landmark 1980 Supreme Court decision in *Diamond v. Chakrabarty*, which established that a genetically modified organism could be patented, opened the floodgates. Soon after, Monsanto and a handful of big chemical and pharmaceutical conglomerates—including DuPont and Pfizer—were investing significant money in applying emerging gene-splicing techniques to crops.

Almost immediately, Monsanto execs directed its new biotech team to engineer crops that could withstand Roundup, so farmers could spray their fields at will, eliminating weeds while crops thrived, reports Daniel Charles in his 2001 book on the birth of ag biotechnology, *Lords of the Harvest: Biotech, Big Money, and the Future of Food.* "Roundup tolerance became the project that bankrolled Monsanto's pursuit of genetically engineered crops," he writes. The idea was a natural. Roundup had emerged as the herbicide of choice in the Midwest, used to "burn down" the weeds in a field before planting and to clear hedgerows and fencerows. A successful Roundup-resistance gene would mean farmers could spray the chemical even after crops sprouted in the field—dramatically expanding the market for the wonder weed killer.

For much of the 1980s, Monsanto scientists struggled mightily to find a gene that would effectively confer Roundup resistance and also allow plants modified with it to grow robustly. After years of near-misses and tantalizing failures in the lab, they "discovered that nature had trumped all of their efforts," Charles reports.

The eureka moment came in 1989. It involved a vast Luling, Louisiana, chemical factory along the Mississippi River, where Monsanto had manufactured Roundup for years. Outside the Luling plant, an uncontrolled experiment in glyphosate exposure had been taking place for years, as Charles writes:

> There are glyphosate residues in the ponds, in the mud at the bottom of the ponds, and in the soil alongside. Those residues exert a steady pressure on the population of microorganisms in the water and the soil, eliminating those that are sensitive to glyphosate and selecting for those that are less vulnerable.
>
> People from the company's cleanup team collected sludge samples from affected ponds for analysis in the early 1980s. The samples sat idle for years in a company lab, until the GMO team thought to look there. When they did, they found the gene they were looking for in that glyphosate-laced sludge from Luling—a gene that "proved to tolerate Roundup far better than any gene the scientists had created in the laboratory," and didn't interfere with plant growth.

In other words, the scientists found that in landscapes subjected to regular Roundup exposure, organisms evolve to resist Roundup. Like plants, pond-dwelling bacteria and fungi synthesize protein with the very enzyme that glyphosate blocks. Constant Roundup lashings kill most of these microscopic creatures, but a few will have genetic mutations that allow them to survive, and they'll pass on the gene to progeny. Such mutant specimens are needles in a haystack, and the company had unwittingly torched the hay, making the needles easy to find.

After years of effort led by Fraley, the R&D team succeeded, focusing their efforts on splicing the Roundup-resistant gene into three of the four most most-planted U.S. crops: corn, soybeans, and cotton. (Monsanto shelved plans for Roundup Ready wheat in 2004.)

This combination—a broad-spectrum herbicide, sprayed on widely planted crops engineered to withstand it—emerged as Monsanto's guiding economic driver, the killer app that delivered the great bulk of those above-mentioned billions of dollars in profits for Monsanto's shareholders. And it was the impetus for the company to shed its industrial-chemical baggage (and its reputation for corporate malfeasance) and rebrand itself as a "life sciences" company, devoted to feeding the world through "sustainable agriculture."

But the turn to agriculture didn't transform the narrative, nor did it convince the company's executives to stop downplaying evidence of harm while promoting dubious products. And Roundup Ready technology has joined DDT and Agent Orange among Monsanto's highly lucrative, deeply problematic business triumphs.

From the start, Monsanto had to grapple with the possibility that its Roundup Ready crops would trigger Roundup-resistant weeds, keeping farmers on the same search for the latest weed killer that they embarked on by embracing 2,4-D after World War II. In its 1993 petition to the U.S. Department of Agriculture to deregulate Roundup Ready soybeans, Monsanto insisted that glyphosate, the active ingredient in Roundup, is "considered to be an herbicide with low risk for weed resistance." Citing agreement from university scientists, the company declared it "highly unlikely" that widespread use of Roundup Ready technology would lead to resistant weeds. At the time, no weed species exhibited the ability to stave off the potent chemical.

The petition made no mention of those ponds outside the factory in Luling, where bacteria regularly doused with Roundup had evolved to do just that.

The company presented its high-tech seeds as a boon to humanity. The Roundup Ready trait would allow farmers to "take advantage of [Roundup's] well known, very favorable environmental characteristics," the petition continued. Farmers employing the technology would no longer have to till the soil to kill weeds during the growing season, which would result in "increased soil moisture, while reducing erosion and fuel use." Relying solely on Roundup all season long would also reduce herbicide use overall, the document suggested.

The USDA accepted the logic, greenlighting Roundup Ready corn, soybean, and cotton crops in the mid-1990s. Farmers throughout the Corn Belt rapidly embraced them, paying the "technology premium" for Roundup Ready seeds so that they could douse their fields with weed killer all season long, delivering "clean" fields and unharmed crops. The company's other mid-1990s triumph was corn and cotton seeds altered with genes from *Bacillus thuringiensis*, known as Bt, a common soil-borne bacterium that's toxic to certain insects but considered harmless to people and animals. Bt spray had been used in organic farming as a natural bug killer for decades before Monsanto tapped its genome. The resulting "Bt crops" essentially produce their own insecticides.

For a while, the technology worked as advertised. Farmers controlled weeds and insects while cutting back on total herbicide and pesticide use; crop yields jumped. In the first several years of Monsanto's genetically modified revolution in the Corn Belt, farmers' biggest problem was the familiar one of low prices due to crop overproduction.

But then things started to unravel. Despite Monsanto's assurances, weeds began to resist Roundup. The mechanism was basic evolution, the force that allowed those pond-dwelling bacteria in Luling, Louisiana, to select for the Roundup-resistant gene. Roundup is an extremely effective weed killer, so the population of weeds that can survive it was unusually low. (Monsanto used this fact to argue that Roundup was highly unlikely to cause resistance.) But apply it over millions of more or less contiguous acres over enough growing seasons, and you'll eventually hit on weeds that can resist it. They'll pass their resistance genes onto offspring, and you've got resistant weeds spreading through the landscape, shaking off the chemical deluge.

At first, Monsanto pretended away the looming resistance crisis and convinced farmers to do so as well. In a 1997 paper published in the academic journal *Weed Technology*, a team of four Monsanto scientists echoed the USDA petition from five years before, insisting that because of Roundup's "unique" chemical structure, weeds were "unlikely" to develop resistance "under normal field conditions."

By 2004, Monsanto was placing advertorials in Corn Belt trade journals assuring farmers that resistance was essentially impossible. One ad pointed to a study on small test plots in Nebraska, which found no resistant weeds despite seven years of continuous Roundup use. At the time, the Iowa State University weed scientist Bob Hartzler found the study unconvincing; in his role as an extension agent (basically, a translator between agriculture researchers and workaday farmers), he let his readers know it. "Using small plots to conclude there is little risk of resistance with systems relying solely on glyphosate can be compared to buying scratch and win tickets in the Iowa Lottery," he wrote in his newsletter. He added: "Using the results of a few small field trials to conclude that continuous use of glyphosate poses little threat of selecting resistant weeds is similar to accusing the Iowa Lottery of running a crooked game because you've never gotten the winning ticket." Hartzler urged farmers to find other ways to "utilize integrated management systems that rely on a variety of weed control tactics," not just Roundup.

Hartzler's suspicions were confirmed that same year, when a pesky weed called waterhemp showed the ability to withstand Roundup in a Missouri soybean field—the first glyphosate-resistant waterhemp species recorded in the United States. "By 2006 or 2007, more than half of our counties had glyphosate-resistant waterhemp," the University of Missouri weed scientist Kevin Bradley told a trade journal in 2018. Also in 2004, a team of USDA scientists came out with a paper documenting eight weed species worldwide that had evolved to resist glyphosate, including two in the United States (one of which was waterhemp). "The problem of weed species evolving resistance is expected to increase with intense use of glyphosate and continued adoption of GR [glyphosate-resistant] crops without rotation with non GR crops," the scientists wrote.

Ironically, this new threat, too, ended up benefiting Monsanto. To fight resistant weeds, farmers both boosted their doses of Roundup and resorted to showering their fields with a cocktail of older, more toxic herbicides—most of which were also made by Monsanto and its peers. As a result, Roundup use in the Corn Belt continued to rise steadily, and the number of Roundup Ready corn and soybean acres kept climbing. By 2011, glyphosate was so widely used that U.S. Geological Survey researchers routinely found traces of it in air, rain, and stream samples taken in Iowa and Mississippi (another state with large acreage of Roundup Ready cotton, soybeans, and corn).

In retrospect, the rise of Roundup Ready crops stands as a bellwether event in the history of U.S. farming. Along with nitrogen fertilizer and tractors, they are the third of what I think of as the trifecta of twentieth-century innovations that have transformed the Corn Belt's landscape, setting the parameters for our most important growing region as we head into the twenty-first century's third decade. Farmers came to rely on the "clean," herbicide-sprayed fields that Roundup Ready crops made possible. At the same time, they found themselves locked into a kind of permanent chemical war against ever-evolving weeds, with Monsanto and its peers playing the role of defense contractors.

As the threat of glyphosate-resistant weeds became clear, Monsanto and its peers merely rolled out crops engineered to resist both glyphosate and older, more toxic chemicals. The chemical that started the herbicide revolution, 2,4-D, had seen its star dimmed by the rise of glyphosate, as had a related blockbuster herbicide from an earlier era, dicamba. Rather than invest resources in discovering new compounds to fight weeds, the industry reverted to its past. Encouraged by Monsanto, the big agrichemical-seed firms explicitly saw the rise of glyphosate resistance as an opportunity to sell more herbicides. As *Bloomberg Businessweek*'s Jack Kaskey reported in 2011:

> Monsanto Chief Executive Officer Hugh Grant says competitors' efforts to develop their own herbicide-tolerant crops isn't a threat to the company's flagship business. Seed companies will cross-license each other's genetics to create crops able to withstand multiple weedkillers, he says, and spraying fields with a mix of herbicides will kill the superweeds and give Roundup Ready crops new life. Monsanto itself is adding resistance to dicamba, an older weedkiller, to Roundup Ready crops for sale by 2015. "The cavalry is coming," Grant says.

By 2012, Monsanto and Dow AgroSciences had both petitioned the USDA to start the years-long process of approving the multi-resistant crops. Again, the companies had to push back against concern that the novel products would cause weeds to develop resistance. In its USDA petition for soybeans that could withstand not just glyphosate but also dicamba, Monsanto declared it "unlikely" that weeds would develop resistance to the dicamba-Roundup cocktail. It even added that the product could "potentially delay or prevent the development of dicamba- and glyphosate-resistant weeds as well as resistance to other soybean herbicides in U.S. soybean fields."

In its brief push for a rival glyphosate-resistant soybean that could also withstand 2,4-D, Dow AgroSciences claimed, boldly, that use of its soybean product would "allow growers to proactively manage weed populations while avoiding adverse population shifts of troublesome weeds or the development of resistance" and "should not result in 2,4-D resistant weeds becoming significant issue in soybeans."

Dow and Monsanto both reasoned that it would be difficult for weeds to survive two different herbicides that attack them simultaneously in entirely different ways. A 2012 paper by a team of researchers led by Penn

State University weed scientist David Mortensen shredded that argument. The team noted that resistance to two or more herbicides isn't a rare occurrence at all: globally, no fewer than thirty-eight weed species across twelve families had already evolved resistance to two or more herbicides—"with 44% of these having appeared since 2005." They added that on millions of acres of farmland in the Midwest and South, many weeds will need to develop only a single resistance pathway, because they're already resistant to Roundup. That is, when farmers apply 2,4-D at will to weeds that are already resistant to Roundup, they'll essentially be selecting for weeds that can resist both.

All in all, the authors concluded, chances were "actually quite high" that the new products would unleash a new generation of superweeds that could shake off Roundup and either 2,4-D or dicamba. In that case, farmers would likely respond just as they had responded to the advent of Roundup resistance—by applying ever higher doses.

In the end, amid much pushback from environmental groups, U.S. regulators took the companies' side and approved the products. The chemical cavalry, to quote former Monsanto CEO Grant, galloped onto the scene in 2016 and 2017, when Monsanto (now Bayer) rolled out soybeans that could resist glyphosate and dicamba; and Dow (now Corteva) debuted a rival product that could withstand glyphosate and 2,4-D. Just as weeds that could shake off Roundup started showing up soon after Roundup Ready crops hit fields, weeds that can resist dicamba, 2,4-D, and glyphosate had already turned up in a Kansas field by 2019.

The rapid evolution of resistance wasn't even the biggest scandal surrounding the debut of Monsanto's dicamba-ready soybeans. As early as 2012, many observers noted that dicamba is highly volatile, especially in warm months. That is, after it has been sprayed on, it's prone to convert into a gas and be carried on the wind to nearby fields. Monsanto had an answer: it had conjured up a "low volatility" version of dicamba that, it claimed, when used properly, should inspire "no concern of off-site movement due to volatility."

Again, experience in the field trumped the company's spin. Every year since its debut in 2016, millions of acres of soybeans, vineyards, home gardens, and oak forests have been hit by dicamba drift. In October 2018, company executives hailed the debut as a "tremendous success." That year, the dicamba trait had conquered 43 percent of the U.S. soybean crop.

As Monsanto's herbicides and multi-resistant crops continued to saturate the marketplace, the company associated with DDT, PCBs, and Agent Orange faced a familiar specter: public outrage. Much of the company's early PR troubles were related to consumer fears over genetically engineered crops, which had gone from zero to ubiquitous in just a decade. But around 2010, the dangers of herbicide-resistant weeds started making headlines, and not just in specialized media but in big establishment outlets such as NPR, the *New York Times*, and the *Wall Street Journal*. Observers began to voice very real concerns that genetically engineered crops had essentially become a marketing tool for the sale of the real profit maker: herbicides.

In 2012, a ballot initiative in California proposed requiring all foods containing GM ingredients to be labeled as such. This would have required tags on nearly every item in the supermarket, since fats, sweeteners, and thickeners derived from corn and soybeans permeate processed food. Monsanto and its peers spent tens of millions of dollars fighting back the California proposition, but other states began floating their own labeling proposals. In a 2013 poll by the public-opinion firm Harris, measuring the "reputation quotient" of the "most visible" U.S. companies, Monsanto landed in forty-seventh place out of sixty firms, joining Wall Street banks and cable-TV companies in the least-popular tier.

That same year, Monsanto launched a charm offensive to destigmatize GMOs in the court of public opinion. The company has "been absolutely riveted and focused on giving technology and tools to farmers to improve their productivity and yield, and we haven't spent nearly the time we have needed to on talking to consumers and talking to social media and really intercepting this" opposition to biotechnology, Fraley told reporters on a visit to *Politico*'s Washington, D.C., editorial offices in 2013. Months before Fraley made the rounds of New York City and D.C. newsrooms to take the case for Monsanto directly to the press, the company had hired press agents

FleishmanHillard to reshape its reputation. FleishmanHillard had served a previous stint as Monsanto's main PR firm in the 1980s, and was the shop of choice for the tobacco industry during the secondhand-smoke controversies in the 1980s and '90s.

By 2014, FleishmanHillard-advised Monsanto was getting up to PR stunts. In one instance, it quietly paid "mommy bloggers" to attend panels on the wonders of GMOs, in hopes they'd bring the good news to their readers. In an embarrassing episode, Monsanto worked with the magazine publisher Condé Nast to try to lure high-profile industry critics such as Marion Nestle, Michael Pollan, and Anna Lappé to appear in company-sponsored videos, by offering as much as $25,000 in compensation and downplaying the company's name in the invitations. The link ultimately came to light, and all the critics declined.

Simultaneously, Monsanto was secretly working with another PR firm, Ketchum, to coordinate and promote pro-GMO messages among university researchers, as emails uncovered by the nonprofit group U.S. Right to Know and *New York Times* reporter Eric Lipton showed. "Monsanto, the world's largest seed company, and its industry partners retooled their lobbying and public relations strategy to spotlight a rarefied group of advocates: academics, brought in for the gloss of impartiality and weight of authority that come with a professor's pedigree," Lipton wrote in a 2015 exposé.

Perhaps its most successful PR strategy from this era was inviting influential critics to tour the company's St. Louis research facility and speak at length with Fraley. Before Bill Nye's visit to Monsanto offices in 2015, the bow-tied TV personality, best known for his educational Disney Channel show *Bill Nye the Science Guy*, trained a skeptical eye on the industry. Back in 2005, he devoted a nuanced episode of his TV show to GMOs, the takeaway of which was hardly fire-breathing denunciation: "Let's farm responsibly, let's require labels on our foods, and let's carefully test these foods case by case."

In his 2014 book *Undeniable: Evolution and the Science of Creation*, Nye reiterates these points. His concern about GMOs centered mainly on the unintended consequences of growing them over large expanses. He cites the example of Roundup Ready crops, which have been linked decisively to the decline of monarch butterflies. The butterflies rely on abundant milkweeds, which have been largely wiped out in the Midwest by GMO-enabled herbicide use. Nye praised certain GMOs, such as corn engineered to kill insects, but concluded, "If you're asking me, we should stop introducing genes from one species into another," because "we just can't know what will happen to other species in that modified species' ecosystem."

After visiting the Monsanto labs and having an audience with Fraley, Nye's doubts fell away. In a February 2015 video interview filmed backstage on Bill Maher's HBO show, Nye volunteered that he was working on a revision of the GMO section of *Undeniable*. He gave no details, just that he "went to Monsanto and I spent a lot of time with the scientists there." As a result, he added with a grin, "I have revised my outlook, and am very excited about telling the world. When you're in love, you want to tell the world!"

In a 2017 episode of his Netflix show *Bill Nye Saves the World* called "More Food, Less Hype," Nye crystallized his new view. After sending a correspondent to a farmers market to lampoon shoppers' negative views on biotechnology, Nye brought out Fraley to defend GMOs on a panel alongside Julie Kenney, whom he identified as an Iowa corn and soybean farmer. He neglected to mention that Kenney is also a longtime PR professional representing the agribusiness industry—including a nine-year stint as a communications executive at the GMO-pesticide giant DuPont Pioneer (now part of the even bigger firm Corteva). Nye also included a more skeptical voice on his panel, Fred Gould, a professor of entomology at the North Carolina State University. But his conversion to a pro-GMO stance was clear: "I believe the advantages of GMOs outweigh the downsides," he announced in the introduction.

Heading into my meeting with Fraley in 2016, part of me wondered whether it was my turn to be struck like Saul on the way to Damascus, as Nye seemed to have been during his own visit. Monsanto was still riding high, despite its bruises. By that year, more than 90 percent of soybeans and corn were genetically modified to survive being doused with herbicides, and 80 percent of corn tweaked to express Bt proteins. (Most corn contains what the industry calls "stacked traits": both herbicide tolerance and Bt.) Monsanto cleared $6 billion in gross profit from selling corn and soybean seeds and associated GM traits in 2016, and another $943 million selling herbicides. Moreover, this blockbuster cash haul came in the midst of a prolonged slump in the global market for these crops.

Alongside Washington University anthropologist Glenn Stone—whose careful research over decades has mounted serious critiques of biotechnology—I spent five hours winding through the labyrinthine corridors of the vast facility, speaking with researchers, scientists, and managers from all five of the company's "innovation platforms:" biotechnology, plant breeding, soil microbes, pesticides, and data science. Our long march through the building was bookended by interviews at a conference table with Fraley.

In his book *Lords of the Harvest*, Charles portrays Fraley as a ruthless figure. He quotes an anonymous former Monsanto exec likening Fraley to Lavrentiy Beria, the chief of the secret police under Stalin in the USSR. "He's a really smart guy, but absolutely merciless," a former Monsanto exec tells Charles. I found Fraley formidable: a barrel-chested man with a large bald head and a steady, skeptical gaze. But he was friendly and even occasionally lighthearted. We joked about our common baldness.

I braced myself for an onslaught of rhetoric about how the company's genetically altered seeds were vitally necessary to "feed the world" as global population moves toward 9 billion and the planet warms. I expected to hear about wonder crops engineered to thrive amid the coming droughts and floods. But Fraley and his team mostly steered clear of these industry-standard tropes. They gave a fairly subdued view of the matter, focusing instead on their core products: tweaks to the herbicide-resistant and insecticide-carrying crops the company had been selling for more than twenty years.

Indeed, for a corporate warrior who has spent his career promoting and defending GMOs, Fraley delivered a rather sober assessment of them. "When people think of us, they always think of Monsanto as the GMO company," Fraley said. "I helped invent it [GM technology], and we've been the leader in that space. But by far the biggest contribution we've made to yield gains around the world is how we've applied biotechnology to the [classical] breeding engine itself."

To hear Fraley tell it, Monsanto was an old-school seed company that turned to GMO technology only as a last resort. (Never mind that it was a chemical company that barreled into the seed business in order to monetize an herbicide-resistant gene it had discovered.) Gene transfer is an expensive technology. "It costs us one hundred and fifty million dollars to develop a GMO product," he said. For most purposes, good old-fashioned plant breeding, sped up by genomic tools, trumps transgenics, he argued. He insisted that classic breeding was the "mainstay," "base engine," and "core" of Monsanto's business, and stressed that it always would be, adding that it takes up half the company's R&D budget. As for genetic engineering, "We only use it on things we can't do any other way. The only way to get a Bt gene into a corn or soybean plant is to use gene-transfer technology and create a GMO," he said.

I was gobsmacked. Rather than trying to convince me to stop worrying and learn to love GMOs, Fraley was trying to convince me that Monsanto wasn't all that into the controversial technology.

I tried to provoke a bit of boosterism out of him by asking about wonder crops: Couldn't GM technology, as advocates sometimes insist, one day achieve grand visions, like corn that mimics legumes and snatches nitrogen out of the air for self-fertilization? "Not likely," Fraley said. He and his team had concluded that creating a nitrogen-fixing corn through gene transfer would require thirty separate traits, and thus be way too costly. Strikingly, we didn't hear a peep about the crops that had once been claimed as humanity-saving triumphs: corn that grows well in drought conditions, say, or thrives with minimal amounts of nitrogen fertilizer.

What about heralded gene-editing technologies like RNA interference and CRISPR? Rather than plucking genes from one organism and inserting them into another, this approach to gene editing works its magic within the genome of a single target organism, altering DNA at precise locations. It can deactivate, activate, or alter genes.

Here Fraley warmed up. He declared these technologies "transformative." But crucially, he took pains to classify them as "superdirected" versions of classical breeding, not amped-up GMOs. They "let you breed even faster and better, and allow you to do some of the things you can do [with GM technology], but won't let you introduce a new trait," he said.

While the Monsanto team downplayed the role of GMOs in the company's future, they defended its bread-and-butter products. Fraley accompanied us to the biotechnology wing of the research center, the first stop on our tour, where I'd heard vigorous defenses of a trait that Monsanto has been selling since genetically altered crops first hit farm fields in the mid-1990s: the insect-killing gene from the soil bacterium *Bacillus thuringiensis*.

A researcher from India told us about his childhood on a two-acre farm applying insecticides with a backpack sprayer: a hazardous activity made obsolete, he said, by the rise of Monsanto's Bt cotton in India. Then the same researcher launched into the benefits of another crop, a soybean product now taking off in Brazil. It's engineered to contain both the Bt insecticide and the other GM trait that Monsanto has been selling since the 1990s: resistance to glyphosate, the company's flagship herbicide. In other words, during our stop at the biotech wing, we heard about nothing new, but rather about the same two traits Monsanto has been selling for two decades: herbicide tolerance and Bt.

My favorite episode of our trip was our visit to Monsanto's emerging soil microbial unit, which develops supplements meant to boost soil health and produce more robust crops. "You have more microbial cells in you than you have your own cells," the unit's lead researcher explained. "A plant is no different. I guarantee there are more cells [in soil microbes] than there are in plants." And so Monsanto is working diligently to identify and market the "most beneficial" of the microbes, the ones that can help make nutrients more bio-available to crops or crowd out soil-borne pathogens. And just like people can eat yogurt or take probiotic supplements to add beneficial microbes to their gut biomes, farmers can buy microbial seed treatments and sprays to fortify their soil, he said.

Just as I'm not someone who's readily convinced that Big Pharma is going to come up with some magic probiotic mix that transforms human health, I also don't think Big Agrichemical is going to stumble upon and package just the right combo of microbe species for growing robust crops without lots of fertilizers and pesticides. The microbial communities that exist in animal guts and in the soil have evolved over eons. I suspect that diverse diets and crop rotations—not lab-grown potions—are key to engendering healthy biomes, both within our bodies and in the dirt.

Still, I was happy to see Monsanto thinking in terms of adding life to soil, not just dousing it with chemicals designed to kill almost everything in sight. So what I saw next made my jaw drop. The researchers pointed to a glass case featuring hearty-looking corn and soybean plants grown with microbial products already on the market, with placards featuring names like Control, Tag Team, Optimize, and Biological.

But for each of the six products, I noticed, the words "Acceleron® Fungicide and Insecticide" appeared under the product name. I cleared my throat and asked why "biological" products were being marketed under biocide labels. The researcher handled the question in stride. "What we've done is taken biological products and put it on top of the fungicides and insecticides most [corn and soybean] growers are using today," he said.

Eventually, he said, they hope growers will begin to actually replace the chemicals with microbes. (In case they don't, Monsanto seems to be hedging its bets. Earlier in the tour, I had met people from the chemicals division who informed me that the company is also developing new fungicides.)

Later, I looked up the Acceleron product. It turns out it's marketed by Asgrow, one of Monsanto's seed subsidiaries. It's a mix of pyraclostrobin, a potentially worrisome fungicide, and Imidacloprid, a member of the neonicotinoid class of insecticides that's suspected of harming bees, birds, and aquatic creatures.

As for the microbial mix the company mashes up with those potent chemicals: it's made up of *Bacillus amyloliquefaciens*, a common soil bacterium, and *trichoderma virens*, which is, yes, a fungus. So probably the most remarkable thing I learned on my visit is that Monsanto is marketing a fungus and a fungicide in the same package.

Looking back, Fraley was pursuing a familiar strategy with me, one seemingly baked into Monsanto's corporate DNA: defend existing blockbuster products, no matter how problematic, and redefine the mission in more benign terms.

In a $66 billion deal that formally closed in 2018, Monsanto skulked off the historical stage and into the maw of a former rival, German chemical giant Bayer. At the time, each of its two flagship herbicides, Roundup (glyphosate) and dicamba, was engulfed in legal woes. One set of lawsuits concerned the "drift" potential of volatile dicamba particles; the other claimed glyphosate causes cancer and that Monsanto knowingly hid that information from the public (Bayer denies both claims).

Despite its inglorious exit, Monsanto was a rollicking success for its shareholders, and a case study of the agrichemical sector's ability to squeeze enormous profits out of the economically brutal task of farming corn and soybeans at large scale.

At the time of Monsanto's IPO, in October 2000, the market valued it at $5 billion. The Bayer purchase price in 2018 marked a fourteen-fold return for shareholders who bought at the IPO and reinvested dividends. Over the same time period, the overall stock market, as represented by the S&P 500 index, delivered a return less than a third that size. In short, Monsanto made its investors loads of money.

"You can only squeeze so much blood out of a rock," Jonas Oxgaard, an analyst who covers the chemical industry for the brokerage firm Bernstein, told me in March 2019. The topic was those glyphosate lawsuits. A second jury had just sided with a plaintiff on the question of the weed killer's carcinogenic properties and Monsanto's role in covering it up; thousands of similar lawsuits were queued up. In Oxgaard's metaphor, Bayer/Monsanto is the rock, and the blood seekers are the cancer-stricken people suing the company for alleged harms from glyphosate. But the same imagery aptly describes Monsanto's business triumphs, too: the tremendous profits it managed to wring out of the fraught enterprise of farming by selling products of dubious ultimate value.

6

Gully Washers

To see the Corn Belt in its full productive glory, your best bet is to visit in July. At the height of the summer growing season in an average year, you'll see a show of force, a statement of industrial agriculture's raw power. Crops are planted so tightly and grow so briskly that the region generates more photosynthetic activity than any other spot on Earth during July, a 2014 NASA analysis of satellite imagery found.

But to see the region when it's vulnerable—to get the best sense of what could possibly go wrong with such a regimented system—you have to go in the spring, sometime between the first thaw and the end of planting. It's like bumping into a champion heavyweight boxer at the hotel buffet at breakfast the morning after an epic brawl. In this spring interlude, the vast majority of ground is uncovered, save perhaps for stubble left over from the harvest. When the rain hits bare ground, soil begins to move.

Few people know more about the interaction of soil and water—and the sometimes-disastrous consequences that can ensue—than Rick Cruse, a professor of agronomy at Iowa State University. Cruse is the foremost authority on soil erosion in the Corn Belt. Since the early 1990s, he has run the Daily Erosion Project, which estimates soil loss in Iowa and surrounding states. He also runs the Iowa Water Center, a collaboration between Iowa State and the U.S. Geological Survey to study the state's water quality.

In early June 2019, I got Cruse to give me a tour of Iowa farmland around Ames in my rented Hyundai. He's a tall, slender fellow with close-cropped graying hair and a longtime teacher's gentle, patient way of explaining things. Normally by that time of year, I'd have been too late to catch the transitional lull in full display. The landscape's annual green carpet would already be established. The corn would be at least knee-high, and the soybeans not far behind: another bin-busting harvest in its first blush of youth. Not that year.

During the second week of March, what meteorologists call a bomb-cyclone storm rampaged across the plains, bringing blizzards, heavy rains, and monster winds from Colorado to the Great Lakes. Bomb cyclones are fierce, hurricane-like events that form when a region of warm air meets one of cold air, causing a fast drop in barometric pressure. The storm system brought hard rain, which pelted the snow, melting it. Because the soil was still frozen from a long, cold winter, the water couldn't percolate downward, as it would in warmer conditions, at least a little. So the resulting cascade moved to lower ground, triggering historic flooding.

But water wasn't the only substance the March bomb cyclone pushed downhill. When a sheet of water moves across bare ground, it doesn't percolate downward, but it does melt the top layer, turning into "mush," Cruse told me. "If you step in it, it's like walking on pudding." Saturated with water, soil is prone to wash away.

The bomb cyclone on its own would have made for a remarkably destructive spring. "Baby calves were swept into freezing floodwaters, washing up dead along the banks of swollen rivers," the *New York Times* reported from the ground in Nebraska that March. "Farm fields were now lakes." Three people died, and in many places, "rail lines and roads that carry farmers' crops to market were washed away by the rain-gorged rivers that drowned small towns," the *Times* added.

But relentless rain continued through June, leaving nearly the entire Corn Belt a mud pit and delaying planting by at least a month. One farmer located just over the border in Illinois, Brian Corkhill, told me he had experienced "basically six weeks straight of rain" by early June, giving him only two days that were dry enough for planting the entire spring. By that time of year, he told me, he's typically "long since done" sowing 1,300 acres of corn and soybeans. Under normal conditions, "corn would be knee-high or a little taller," he said. Instead, in 2019 he didn't even start planting until May 16, a day after he's usually done. By the first week of June, he had just over half of his corn and none of his soybeans in the ground. Stories like Corkill's resounded through the Corn Belt, from Nebraska to Ohio, Minnesota to Missouri. It was the region's slowest and latest corn-planting season on record.

But it wasn't the lateness of the planting and the disarray of the fields that Cruse was eager to show me when I caught up with him in June. Thanks to the Corn Belt's overwhelmingly industrialized and technology-laden consolidation, the world has plenty of corn and soybeans, and a massive abundance of the foods derived from them: cheap meat, sweeteners, and fats. In the grand scheme of things, one year's short corn and soybean harvest wouldn't have a massive impact on a food system awash in those crops.

Rather, Cruse wanted to show me what the wet, wild spring meant for the region's soil. As we rolled down two-lane highways at 60 miles per hour, we saw field after field of bare ground. Some of the plots had shoots of corn poking through; here and there, other patches were submerged in a foot or two of water, the aftermath of recent rains. The landscape's dominant feature by far: mud.

This flagrant display of naked land emerged as a rite of spring in recent decades. But historically, it's an anomaly. For millennia, the region was dominated by perennial prairie grasses and wildflowers, which plunged their roots deep into the ground and provided a thick stand of vegetation. When heavy spring rains hit, the roots below anchored the ground and the stalks above buffered it, holding soil in place despite the deluge. Moreover, the cycle of growth and decay of root mass and stalks provided a steady helping of organic carbon to the soil, feeding a vast web of microscopic organisms. In turn, these creatures recycled nutrients, making them available to the grasses for each spring's new-growth surge. The grasses didn't just protect the soil from the brute force of hard rain; they also provided a kind of sponge for rainwater to percolate downward, with the roots serving as channels. This percolation effect reduced the frequency of heavy floods and provided a store of water that could maintain plant growth during hot, dry summers.

That was then. Now we're in the world of corn and soybeans, planted in the late spring and harvested in the fall, leaving the ground bare from November through the end of June, when the crop canopy is high and thick enough to protect it.

Every few minutes, Cruse slowed down to show me a sight that clearly pained him: large gashes that follow contours in farm fields, formed by water runoff during heavy rains. Known as "ephemeral gullies," these channels represent soil that's been carried off fields and dispersed into ditches and streams. They develop on sloped land that's contoured in a way that pushes lots of water into a relatively small area, generating enough energy to cut into the land. "The faster water moves, the more soil it will carry," Cruse explained. That's why gullies appear after massive rain events, of the kind that regularly pummeled the land in spring 2019. They're called "ephemeral" because when one forms in the spring, farmers typically till the ground and push fresh soil onto it to cover it over. But that stopgap solution makes the gullies prone to forming again the next year. They're essentially pipelines, periodically filled by farmers, that remove millions of tons of topsoil from prime farmland.

Cruse explained that when he was a kid growing up on a farm in northeast Iowa in the 1960s, his father had a special name for the spring storms that occasionally raged through the region, dropping massive amounts of water over a short time. He called them "gully washers," because of their tendency to expose these vulnerable regions in bare fields. One spring Sunday, a particularly savage storm blew through, and in its aftermath, the elder Cruse gathered the family in the sedan for a drive to survey the damage, calling out the names of farmers who had allowed huge swaths of their fields to be gouged away in the deluge. "My father told me something I'll never forget: 'Soils are connected to *everything*. Without soils we have *nothing,*'" Cruse told me. And he has been surveying the damage to Iowa's fields from storms ever since, documenting and spreading the lessons he learned on that primal Sunday drive.

After that succession of gully washers in spring 2019, those gouges marked the land at frequent intervals throughout our drive. At one point, Cruse stopped to show me a particularly gruesome one zigzagging down through a muddy field of wan corn shoots, terminating in a mud-caked ditch along the road. "There was a gully when I drove past here just after the bomb-cyclone storm," he said. When Cruse returned after a dry spell a few weeks later, the farmer had plowed the field and pushed soil into the void, making it vanish. "But the recent rains have cleaned it out again. So he's had two batches of soil export in one season. It's just disgusting."

The only good thing about ephemeral gullies is that you can easily see them from the road and capture them in photos. They make the problem visible, comprehensible, and relatable, in the way that threats to charismatic megafauna such as elephants and lions help the public understand wide-scale species loss. But when water and soil meet, there's another, subtler, more devastating way they can interact. What agronomists call "sheet-and-rill erosion" occurs when raindrops hit bare ground, dislodging soil particles and sending them flowing down gently sloped hills. As the water flows, it picks up more particles and cuts tiny channels in the soil surface called rills. (When this activity is concentrated by contours in the land into a tight-enough area, it forms a gully.) Whereas gully erosion bears away soil in a spectacular event, sheet-and-rill is a slow-drip phenomenon.

When it has gone on long enough, though, the trained eye can spot severe erosion the landscape. We pulled over beside an unremarkable-looking spot. There was no gnarly gulley, just a bare field with a gentle upward slope. "See the color change on the hills as you look up?" Cruse asked. I did. It went from a rich black toward the bottom to a washed-out brown. The upper part was lighter in color, he said, because the rich black earth had eroded away, leaving behind less fertile subsoil, which is much lower in microbiological activity, organic matter, and fertility. Then he noted another feature: "You can see rocks on this hill. Know why they're there?" I did not. "Rocks exist throughout—most of them are buried in the soil." He said it's common wisdom among farmers that rocks migrate upward through the soil profile, eventually reaching the soil surface. "That's not true. During rainfall events, rocks don't move, but soil does." As soil vanishes, he said, the rocks stay in place and become exposed. "Rocks are a sign that soil is moving."

When we got back on the road, I asked Cruse to talk me through the math. Just how much soil is Iowa and the greater Corn Belt hemorrhaging, and what is the natural replacement rate? The latter is a key question. Soil is a complex matrix of minerals (45 percent) and organic matter (around 5 percent), with the rest of its bulk made up of water and air. It regenerates slowly, as minerals are released from the underlying bedrock (known as parent material) in a process called weathering.

The precise rate at which soil renews is hazy and difficult to nail down, varying by soil type and climate. In its reckoning, the U.S. Department of Agriculture sees 5 tons per acre as the "magic number" for Iowa, Cruse says. The agency assumes Iowa soils regenerate at that rate—that is, farms can lose up to 5 tons per acre of ground per year without trouble, because natural replenishment will balance the loss. So the prevailing view has been that "if we can limit erosion to five tons per acre, we can do this forever," Cruse said. The trouble, though, is that the USDA has never delivered solid science to back up the 5-ton assumption. In a 1998 paper, Cruse and a coauthor concluded that "seldom has such an important policy been based on such a dearth of defendable data." And the research that has emerged since suggests that the so-called "soil-loss tolerance" rate is actually much lower. In fact, lower by an order of magnitude: around 0.5 tons per acre.

If farmland has about a half ton per acre to spare each year, I asked, how much are Iowa's farm's losing? Cruse's Daily Erosion Project delivers estimates of soil loss in Iowa and surrounding states based on frequently crunched rainfall data, topography, and farm practices. Over the decade, the DEP estimated, Iowa's soil has eroded at an average rate of 5.4 tons per acre annually. There's a catch, though. The data the DEP relies on can paint a reasonably accurate picture of sheet-and-rill erosion, but a precise way of reckoning for ephemeral gullies has proved elusive (though Cruse said his team was getting close). For the time being, the official calculations leave gullies out. However, Cruse's best estimate, based on his own work and that of the research literature, is that ephemeral gullies claim an additional 3 tons per acre annually. Add that to the sheet-and-rill number, and you get an average of 8.4 tons of prime prairie soils washed away per acre per year. That suggests that Iowa—and much of the surrounding Corn Belt land—is losing soil at a rate nearly seventeen times the pace of natural replenishment. And these long-term averages, Cruse stressed, by definition undershoot the damage done during massive soil-loss years like 2019.

These alarming numbers understate the severity of the issue in another important way. Averages paper over huge regional differences. The Des Moines Lobe territory of north-central Iowa—the "prairie pothole" region of drained wetlands that covers about a fifth of the state—is relatively flat and has low rates of sheet-and-rill erosion (though it's prone to gullies). So it pulls down the overall average. Soils in the state's eastern and western parts erode at much higher rates, Cruse said.

As I took that in, we stopped to gape at yet another ephemeral gully, a long, curving scar carved into the land, an abyss where some of the world's best soil once grew crops. "Why?" Rick muttered, pained. "Most of the scenes we've seen in these fields are recurring, year after year after year."

"The soil we're losing is really important," Cruse said, because the stuff at the top is the best: highest in organic matter, nutrients, and biotic life. Carbon-rich topsoil is porous, leaving less water to run off and also creating a store for crops in the coming dry months. That's vitally important during droughts. Iowa's farms aren't irrigated, the way California's are; they rely entirely on rainfall. When summer rains don't come, rainwater captured by soil in the spring months becomes vitally important. In the 2012 growing season, the Corn Belt endured one of its hottest, driest patches ever, and corn yields plunged by more than a quarter, while soybean yields dropped by around 9 percent, compared to pre-drought USDA expectations.

At a certain point, erosion feeds on itself, Cruse said. As eroded soils lose their ability to absorb water, they leave more to run off during rain events, in turn exposing the region to a higher risk of catastrophic flooding—and more erosion.

In the Corn Belt, soil loss on an epic scale occurs when hard rains meet bare ground, Cruse taught me. And hard spring rains have become more frequent in recent years, a trend that will likely continue as climate change proceeds apace.

The chaos of spring 2019 is "not a one-off event" in corn and soybean country, Richard Rood, a meteorologist and professor of climate research at the University of Michigan, told me. Over the past decade, "we've been seeing more and more extreme precipitation in that part of the country," Rood said—not just wetter springs overall but also "more and more precipitation coming in extreme events," like the bomb-cyclone storm that caused severe flooding in March. The Midwest's spring rains are driven mainly by evaporated water from the Gulf of Mexico. As the climate warms, more water evaporates. Warmer air, in turn, can hold more water. With those forces in place, according to Rood, the Midwest is "set up" for more wild springs, in much the same way warmer weather in the Pacific Ocean puts California in the path of vicious winter storms.

Rood's analysis echoes the Fourth National Climate Assessment, a congressionally mandated, quadrennial report by the U.S. Global Change Research Program, designed to assess and summarize the state of science around climate change. The Trump administration did its best to bury the report, quietly releasing it on the Friday after Thanksgiving in 2018. The report found that the Midwest's precipitation patterns have already shifted dramatically: more and more of it comes in the very kind of brief, violent spasms that convulse the soil. To measure increasing storm intensity, the researchers looked at the portion of annual precipitation that falls in high-intensity storms over the course of each year since 1901. The result: the amount of a year's precipitation that occurred in the kind of heavy storms that moved so much soil in 2019 jumped by 42 percent between 1901 and 2016. And the entire jump took place after 1958. That shows just how much today's farmers are encountering a weather regime radically different from that of their grandparents.

The Climate Assessment researchers also looked at what warming trends portend for the next several decades, comparing an optimistic scenario for cuts to global greenhouse gas emissions going forward against a business-as-usual outlook. In the rosy scenario—if the world manages to reduce greenhouse gas emissions over the next twenty years—storm severity in the Corn Belt rises by another 10 to 20 percent by late century. If things continue as they are, we're looking at another 40 percent jump. Both portend many reruns of spring 2019's heavy storms in the coming decades, and thus ever-heavier pressure on the Corn Belt's soil.

As for farming, "increases in humidity in spring through mid-century are expected to increase rainfall, which will increase the potential for soil erosion and further reduce planting-season workdays due to waterlogged soil," the report found. To illustrate the severity of climate effects, the report pointed to the Cedar River Basin in eastern Iowa, where floods considered hundred-year events in the twentieth century are emerging as twenty-five-year events in the twenty-first century.

Recent experience on the ground gives weight to these findings. The spring of 2019 was an impressive example of the kind of weather that can be expected as the climate continues warming, but it was hardly an anomaly in recent years. Indeed, it wasn't even the wettest spring of its own decade.

Between March and May 2019, Iowa's land received on average a mighty 13 inches of rain, causing all the mayhem I saw on my drive with Cruse. But in 2013—the year following a historic drought—the state was hit by 16.6 inches of precipitation over those three months, the highest ever for the period in the state's 141 years of weather recordkeeping. Over two days in late May 2013, researchers from the Environmental Working Group toured farmland around Ames. "On both days they found gullies scarring field after field. Roadside ditches were full of mud and polluted runoff—a very bad sign for Iowa's already polluted streams," the team reported. Two weeks later, Missouri University soil scientist Newell Kitchen told a trade journal that "even the untrained eye of citizens who don't think about agriculture on a day-by-day basis can see the erosion on side slopes." He added, "From the freeway between Kansas City and St. Louis, it doesn't take much to see the scars of erosion that have taken place with the heavy water accumulation."

The Daily Erosion Project's soil-loss map for March–May 2013 looked a lot like the one for 2019. If anything, it showed even more erosion. Large swaths of land lost as much as 32 tons of topsoil per acre—sixty-four times the likely rate of replenishment, and six times the USDA's rate of 5 tons per acre, not counting the gaping gulleys.

The deluges of 2013 and 2019 came on the heels of extremely wet weather in the late aughts. In 2007, Iowa experienced its fourth-wettest year on record. That year, a fierce May storm pounded southwest Iowa, dislodging at least 5 tons of topsoil per acre over a 4.2 million-acre swath (about 16 percent of the state's farmland) *in a single day*—and that's without counting gulleys. A year later, the state experienced its wettest January-to-June period on record, culminating in cataclysmic June floods. "Approximately 1.2 million acres of corn and soybeans were lost, nearly 10 percent of the tillable land in Iowa was underwater, and estimated crop losses surpassed $3 billion," the Congressional Research Service reported. More than 2 million acres of farmland—about 10 percent of Iowa's base—saw erosion of at least 20 tons per acre.

In addition to the ever-more-intense spring downpours that wash away uncovered soil, climate models also suggest continued warming during the summer growing season. "Warm-season temperatures are projected to increase more in the Midwest than any other region of the United States," the Fourth National Climate Assessment reports. So far, the Corn Belt has largely benefited from the warming trend over the past several decades. The time period between the last spring frost and the first fall frost has grown by an average of nine days since 1901—and is expected to grow by another ten days by mid-century. The longer growing season has helped push up yields in recent years, and more carbon dioxide in the air can potentially stimulate plant growth.

But as warming continues, those advantages will be overwhelmed by higher temperatures. A major 2017 study by American and German researchers found that every day above 86 degrees Fahrenheit during the growing season reduces corn and soybean yields by as much as 6 percent in rain-fed farming regions like the Corn Belt. The long, hot days cause soil moisture to evaporate, subjecting plants to water stress. Ongoing erosion compounds the problem. As the ground sheds its carbon-rich, spongy top layers, it loses its ability to gather and retain water for hot days ahead.

In addition to water stress, a hotter Midwestern climate has already "increased the [winter] survival and reproduction of existing insect pests and already is enabling a northward range expansion of new insect pests and crop pathogens into the Midwest," the Fourth National Climate Assessment warns. "Rising humidity also leads to longer dew periods and high moisture conditions that favor many agricultural pests and pathogens for both growing plants and stored grain." In other words, good news for the handful of companies that dominate the pesticide market.

When heavy rain meets bare ground, it's not just soil that goes on the move. It's also the annual deluge of fertilizers, pesticides, and hog manure that farmers apply to it.

Yes, hog manure. In addition to its 24 million acres of farmland, Iowa houses a herd of 23 million hogs, and growing. But as I noted earlier, the role of hogs has shifted dramatically over the past several decades. In 1950, 85 percent of Iowa's more than 200,000 farms kept hogs as a means of transforming excess crops and waste into potent fertilizer and an income stream. At that point, the state's hog herd stood at 10.7 million. By the mid-1960s, as industrial-style agriculture tightened its grip over the land, farmers began to specialize, either scaling up their hog operations or abandoning them to grow more corn and soybeans. The move toward specialization occurred within a broader pattern of consolidation: smaller farms shutting down, and their land being incorporated into larger operations. By 1974, Iowa was down to 117,000 farms, less than half of which kept a total of 11.4 million hogs. The trend of ever-fewer farmers raising ever more pigs has continued apace. In 1998, less than a quarter of the state's 96,700 farms kept hogs—and the ones that did surged in size.

The 23 million hogs in Iowa as of 2018 represent nearly one third of the entire U.S. hog herd, though Iowa accounts for less than 2 percent of the total U.S. landmass. And there are more than seven hogs for every one of the state's three million people. Meanwhile, the average size of a hog operation rose by a factor of more than four between the late 1990s and the late 2010s.

The key question: Where does all the hogshit go? Because there's a *lot* of it. It turns out that when you concentrate tens of millions of hogs in a rainy region traversed by a mighty river, some portion of their excrement is inevitably going to move, fouling ecosystems from local lakes to one of the globe's great gulfs. In short, it's impossible to understand Iowa's water-quality crisis without (figuratively) diving into its vast annual cascade of hog manure. Chris Jones, a research engineer who specializes in water quality at the University of Iowa, crunched the numbers in a series of widely discussed posts during the rain-lashed spring of 2019.

The life of a hog on one of Iowa's farms is short and marked by startlingly rapid growth, Jones reported: "A pig weighs about 3 pounds at birth and about 250 pounds at slaughter a mere 6 months later, so it is gaining more than one pound per day. By comparison, a human infant gains a pound about every 20 days." So it's not surprising that a single pig "excretes 3 times as much nitrogen, 5 times as much phosphorus, and 3.5 times as much solid matter" as a person. All told, Iowa's hog herd delivers as much waste as 83.7 million humans, Jones calculates. So in fecal terms, there are *28* hogs, not 7, for every human Iowa resident.

It should be noted that Iowa's corn and soybeans don't just flow into hog confinements to produce pork. The state also ranks as the nation's leading egg producer (56 million hens) and eleventh-biggest dairy producer (220,000 cows), and it is also a significant producer of calves for the beef market. Add it all up, and Iowa boasts what Jones calls a "fecal equivalent population" of 168 million—the combined human populations of the world's eleven biggest cities, Shanghai to São Paulo; that's 55 "fecal equivalents" for every actual person in Iowa. In terms of density, "in Iowa we are generating as much fecal waste in every square mile as 2,979 people," Jones calculated. That's the highest fecal-equivalent density of any state, twice as much as dairy-intensive Wisconsin.

Livestock manure is, of course, an extremely valuable resource for farming. Biological digestion rapidly recycles plant matter into organic matter and nutrients that can feed soil microbes and crops. Animals have served this role since the dawn of farming twelve thousand years ago, millennia before the nineteenth-century German chemist Justus von Liebig established that the chemical elements of nitrogen (N), phosphorus (P), and potassium (K) are essential to plant growth. There's a general consensus among agroecologists that the most sustainable forms of agriculture include the managed mixing of animals and crops. In his landmark 1944 book, *An Agricultural Testament*, the pioneering British plant pathologist Albert Howard (who is credited with helping launch the organic-farming movement) put it like this:

> Mother earth never attempts to farm without live stock; she always raises mixed crops; great pains are taken to preserve the soil and to prevent erosion; the mixed vegetable and animal wastes are converted into humus; there is no waste; [and] the processes of growth and the processes of decay balance one another.

So it's not insane, per se, that Iowa's farmers use the annual fecal output generated by the state's teeming animal factories to feed crops. The problem, Jones shows, is that manure production on these hog, hen, and cow operations is far too intensive to be absorbed properly by nearby farmland—especially by farmland that is generally left way too vulnerable all spring to massive soil-washing storms. Paradoxically, because Iowa's hog farms tend to be concentrated in the north-central and western regions of the state, and it's expensive to move manure far from where it's made to where it's needed, farmers in some regions are burdened with too much shit, while those in other less-hog-intensive regions don't have ready access to enough. So they apply titanic amounts of industrially produced nitrogen and phosphorus fertilizers—which are also quite vulnerable to escaping during spring rains. In short, Iowa's farmers apply way too much plant food, both in the form of manure and industrial fertilizers, for the land to soak up.

The problem was apparent to some observers as soon as livestock farming began to intensify and become separate from crop farming in the 1960s. In his great 1977 critique of industrial agriculture, *The Unsettling of America*, the renowned farmer-writer-activist Wendell Berry writes:

> Once plants and animals were raised together on the same farms—which therefore neither produced unmanageable surpluses of manure, to be wasted and to pollute the water supply, nor depended on such quantities of commercial fertilizer. The genius of American farm experts is very well demonstrated here: they can take a solution and divide it neatly into two problems.

Current trends point to an intensification of the overfertilization problem. Iowa's hog herd is still growing steadily as the handful of companies that dominate U.S. pork processing look to Iowa, with its abundance of cheap corn and soybeans and its lax environmental laws, to supply pork to a growing global middle class that's demanding more meat. The hypercompetitive nature of Corn Belt farming plays a part, too. As we've seen, farmers tend to respond to the volatile market price for corn by overproducing; the steady pressure to maximize yields pushes farmers to routinely overapply nitrogen, as does the physiology of plants themselves. According to Peter Vitousek, a professor of biology at Stanford University and a leading scholar on the nitrogen cycle, under optimum conditions and using best practices, plants take up only "50 or at best 60 percent" of the nitrogen laid on by farmers. So why do farmers apply so much fertilizer if so much of it is going to waste? Vitousek explains that plants take up different amounts of nitrogen at different points in the growing cycle. To ensure that crops have sufficient nitrogen when they need it most, farmers essentially have to overapply it.

Prices for nitrogen fertilizer, which is synthesized with natural gas, rose steadily between 1996 and 2006 in lockstep with U.S. fossil fuel prices. But the ongoing fracking boom, which began in the mid-2000s, created a spike in U.S. natural gas production and ultimately lowered the price of nitrogen fertilizer, making it a relatively cheap way for farmers to ensure maximum yield. As for hog manure—which is loaded with nitrogen and also phosphorus, and is plentiful in much of Iowa—overapplying it provides farmers cheap assurance that their crops will have plenty of available nitrogen when it's necessary.

Of course, what makes sense for each individual farmer, locked in a battle royale with peers, doesn't make sense for the broader society. Markets are glutted with corn and soybeans and all the foodstuffs they're transformed into. And soils glutted with fertilizers and exposed to erosion cause trouble downstream.

Not surprisingly, when soil loaded with fertilizers and manure exits farms, nitrogen and phosphorus flow away, too: into aquifers, creeks, streams, rivers, wells, municipal water intakes, and lakes. And those key plant foods don't stop stimulating plant growth just because they've exited farms. Algae, too, need nitrogen and phosphorus to grow. Downstream of the Corn Belt, they find these elements in abundance.

Ultimately, most of these chemical building blocks wind up in the Mississippi River, which drains 40 percent of the continental United States—1.2 million square miles in total—including most of the Corn Belt. But the river isn't the only lucky recipient. There's a farm-intensive swath of land west of Toledo, Ohio, that drains into Lake Erie. One day in August 2014, the five hundred thousand residents who rely on Toledo's municipal water utility received a stark warning from city officials: don't drink your tap water, don't wash the dishes in it, and don't bathe your kids in it—even after boiling, which would just concentrate its toxins.

Predictably, the notice caused a run on bottled water. Toledans "waited in lines at fire stations for bottled water, crossed state lines in search of stores with supplies after local outlets ran dry, and drove to friends' homes miles away to fill containers," the *New York Times* reported.

The culprit was a vast algae bloom in Lake Erie, from which the city draws its water. Freshwater growths like Lake Erie's are known as "harmful algae blooms," because they typically produce a compound called microcystin, a toxin that can cause nausea, vomiting, diarrhea, severe headaches, fever, and liver damage. On contact with skin, microcystin can trigger rashes, hives, and skin blisters. It's so caustic that the Iowa Department of Health warns that "breathing airborne droplets containing the toxins, such as during boating or waterskiing" can cause damage. A particularly noxious chunk of algae floated over Toledo's water intake equipment in Erie that August, causing a microcystin spike and the warning not to drink the water, which lasted for three hectic days.

Before the crisis, the city was spending as much as $4 million per year to keep microcystin levels down. Since then, Toledo ratepayers have shelled out $54 million in a new ozone-treatment system the city says will keep the water safe, set to be completed in 2022.

Toledo's water troubles are a throwback to its postwar heyday, when the Rust Belt's booming factories flushed phosphorus-laced wastewater into streams that made their way into Lake Erie, feeding algae growths that rivaled today's in size. But after the decline of heavy industry and the advent of the Clean Water Act in 1972, that source of phosphorus faded during the 1980s, and the infamously polluted lake began to recover. But starting in the early 2000s, massive summer blooms began to reappear.

Toledo is situated at Erie's western edge, right where the Maumee River flows into the lake. The Maumee is surrounded by 5 million acres of land dominated by corn and soybean crops, and the phosphorus the river delivers to Erie each spring has been shown to be the factor that drives the lake's summer blooms. Researchers have documented a jump in the river's phosphorus load that began in the early 2000s and has held steady ever since. But the Maumee River watershed, which extends west into Indiana and north into Michigan, has been part of the larger Corn Belt since U.S. settlers first took over the area. Why did the region's farms suddenly start leaking so much phosphorus and triggering harmful algae blooms in Lake Erie?

One explanation is climate change. Warmer temperatures mean warmer water—which favors the growth of the kind of algae that now spreads in Erie every summer. And since around 1900, spring rainfall on the Maumee watershed has increased by approximately 25 percent, and "the size and intensity of these events have also increased," a team of USDA researchers reported in 2015. More storms means more runoff and bigger and more severe blooms. If both trends continue—warmer water, and more frequent and more powerful spring storms—harmful algal blooms including Erie's will increase in size and duration by mid-century, a team of Tufts, EPA, and Massachusetts Institute of Technology researchers showed in a 2017 paper.

But wetter springs can only partially account for the phosphorus spike. Another factor is the way farmers manage the region's hydrology. For thousands of years, the Maumee watershed was a network of marshland and forested swamps, the remnants of a lake formed by the retreat of the Wisconsin glacier fourteen thousand years ago. A variety of American Indian groups occupied the region over the millennia. The U.S. government seized the area by force in 1794, deeming it the Great Black Swamp, and eventually it was swept up in the land rush that gave rise to the modern Corn Belt. Settlers clear-cut forests and drained swamps and marshlands, exposing rich, black soil suitable for grain farming.

As in the Des Moines Lobe region of north-central Iowa—also a former wetland formed by the retreat of the Wisconsin glacier—farmers in the Maumee watershed beat back the swamp by maintaining belowground drainage pipes known as tiles. These efficiently leach excess water—and chemicals—from the soil and send it into ditches and streams, which flow into the river and ultimately into Erie. Farmers in the region have been maintaining and improving tile drainage ever since. It would be impossible to intensively farm a wetland without tile drains, but they act like sieves, rapidly transferring farm chemicals into water, especially during heavy spring rain events. In a 2017 paper, a group of U.K. and U.S. researchers analyzed weather and stream data and found that more-intense spring storms and recent investments in more-efficient tile drains are two major factors in the upsurge of Erie algae blooms since 2000.

A third factor is likely the expansion of industrial-scale livestock farming in the region, mostly poultry and hogs. In a 2019 investigation, the Environmental Working Group and Environmental Law and Policy Center analyzed aerial imaging and satellite mapping to show that between 2005 and 2018, the number of large livestock operations in the Maumee watershed jumped by more than 40 percent, and the amount of manure they generated increased from 3.9 million tons in 2005 to 5.5 million tons in 2018. Applying that annual cascade of phosphorus-rich manure to tiled farmland provides a gaping opportunity for leakage into the Maumee.

The Clean Water Act won't save Erie this time. It restricts pollution only from "point" sources like wastewater treatment plants and factories; it does not regulate "nonpoint" sources, such as fertilizer runoff from farm fields. Restoring around 10 percent of the land to wetlands would trap nutrients and filter farm runoff, potentially preventing future blooms, researchers say. But without regulatory bite, a push for wetlands restoration is unlikely to succeed. "Among the descendants of the settlers who conquered the Black Swamp, drainage is viewed as sacred, while wetland restoration borders on the profane," the environmental writer Sharon Levy reported in 2017.

Ohio agribusiness interests are actively battling back attempts to regulate farming in the watershed. In 2019, Toledo voters approved an innovative proposal called the Lake Erie Bill of Rights, which confers upon the lake and its watershed the right "to exist, flourish, and naturally evolve." The move, which grew out of a burgeoning "rights of nature" movement with roots in Ecuador and Colombia, is unlikely on its own to force changes in the region's agriculture, and is at any rate being challenged in court by the Ohio Farm Bureau. But it could galvanize a social movement to push back against the annual fouling of the lake.

Every July, the National Oceanic and Atmospheric Administration and a host of agencies within Ohio go through what has emerged as an early-summer rite: predicting the size and severity of the bloom, which reaches full flower in August. Interestingly, given 2019's relentlessly wet spring, that year Erie received 24 percent less bioavailable phosphorus than would have been expected from near-record flows from the Maumee River into the lake, Laura Johnson, director of the National Center for Water Quality Research at Ohio's Heidelberg University, told reporters at a NOAA press conference.

The full amount would have triggered a potentially much bigger algae bloom. The reason for the shortfall largely had to do with 2018's unusually wet fall. Massive rains in October and November delayed the corn and soybean harvest, and prevented farmers from making their usual postharvest fertilizer and manure applications before the ground froze. So the autumn's climate chaos buffered the effects of the spring's climate chaos. Even so, NOAA projected, the 2019 bloom would be formidable—likely bigger than the one in 2014 that triggered the water crisis.

Lake Erie is only the most spectacular of the toxic algae blooms appearing in ponds across the Corn Belt. Toxic algal blooms don't have a big effect on drinking water in Iowa, where most of the supply comes from rivers and wells, which don't develop blooms. But the state has seen a "dramatic" rise in lake beach closings due to harmful algae blooms since 2006, the Iowa Environmental Council reports. And for every don't-swim warning issued because of the presence of microcystin in lakes—which reached thirty-seven in 2016—the state of Iowa makes three or four warnings based on the presence of *E. coli*, a direct consequence of manure overapplication.

In July 2017, I visited Pine Lake State Park, a rare forested area among the cornfields and livestock barns in Hardin County, one of Iowa's most hog-intense regions. I was there with Nick Schutte, a Hardin-based farmer and activist with Iowa Citizens for Community Improvement, which organizes for clean water against the dominance of agribusiness over the state. Schutte told me he'd been fishing and swimming in the lake since childhood. He wanted me to see what had become of it. We met a pair of local residents who are actively trying to improve the lake, and parked and entered a trail through the forest. By the lake, a sign on the Works Progress Administration–era bathhouse confronted us: CONCENTRATIONS OF *E. COLI* BACTERIA AND/OR TOXINS FROM BLUE-GREEN ALGAE EXCEED ACCEPTABLE GUIDELINES FOR RECREATIONAL USE. The sign urged visitors to keep kids and pets well clear of the water, and advised boaters to avoid green patches. CLEAN FISH WELL AND DISCARD GUTS, it added.

When we got to the water, I saw what the fuss was about. In the distance, the opposite shore was inviting, with water gently glistening against a backdrop of gorgeous oaks. The shoreline, though, was empty of people, as was ours. The beach at our feet was grotesque: water the color of pea soup, the bright green broken up by buggy orange scum. A breeze carried a fetid, lightly fecal odor. This place, Schutte said, had been a refuge in his time as a farm kid in the 1990s amid the surrounding sea of corn, soybeans, and hogs.

What about the bulk of the fertilizer and manure runoff that gets borne away by the Mississippi? Ultimately, it makes its way to the Gulf of Mexico, where it wreaks havoc.

The Gulf is a magnificent resource: a kind of natural engine for the production of wild, highly nutritious foodstuff. Nearly three-quarters of wild shrimp harvested in the United States call it home. It is a major breeding ground for some of the globe's most prized and endangered fish, including bluefin tuna, snapper, and grouper. As the climate continues to warm and population grows, sources of low-input, top-quality food like wild seafood will be increasingly precious.

Unfortunately, we in the United States treat the gulf as a kind of sacrifice zone. Nearly a fifth of U.S. oil production comes from the Gulf of Mexico seafloor; and about half of all U.S. oil and natural gas refineries can be found along its shoreline. British Petroleum's long 2010 oil spill there is only the most spectacular ecological blow the fossil fuel industry has delivered there. Other costs include the routine leaking of toxic oil into water and the rapid destruction of coastal wetlands, which are both crucial nurseries for the aquatic ecosystem and buffers for the massive storms that blow in from the south, imperiling New Orleans and other population centers.

Corn Belt agriculture delivers a devastating but less talked-about annual shock to the gulf. Every year, as fertilizer builds at the mouth of the Mississippi River beneath New Orleans, a set of connected algae blooms expands west, typically breaching the Texas border and reaching Houston. When the algae blooms die and decay, they tie up oxygen from the water underneath, creating a condition known as hypoxia, a low-oxygen state inhospitable to sea life. "Habitats that would normally be teeming with life become, essentially, biological deserts," according to the NOAA. Coastal hypoxic zones are a growing global phenomenon, driven by more frequent heavy storms and the spread of U.S.-style industrial-scale agriculture. But the gulf is the world's second-largest such zone, bested only by the one that forms annually in the Baltic Sea.

The dead zone fluctuates in size year to year, based on weather patterns in the Corn Belt. Generally, wetter springs mean more runoff and bigger effects downstream. In its leanest years, the Gulf of Mexico dead zone covers just over 2,000 square miles—a little bit bigger than Delaware. In its most prodigious years (for example, in 2017), it blots out sea life over 8,000 square miles, comparable to the size of Massachusetts. Its average over the thirty-three years between 1985, when researchers first started tracking it, and 2016 is 5,460 square miles: about halfway between the size of Connecticut and New Jersey.

The agribusiness industry and its allies downplay the Corn Belt's contribution to the nutrient load that feeds the dead zone. The industry-aligned Iowa Farm Bureau informed its members in a 2017 blog post that agriculture runoff is just one among many factors feeding the annual bloom, along with "urban run-off," "sewage treatment plant discharge," "fertilizers from golf courses," and "atmospheric nitrogen deposits." In reality, agriculture is by far the dominant contributor. According to the USGS, nutrient runoff from row crops and confined livestock operations delivers 60 percent of the nitrogen load entering the gulf. Leaving aside natural nitrogen deposition from the atmosphere, farms contribute 80 percent of the human-induced nitrogen feeding the dead zone. Golf courses don't come close.

In a 2018 paper published in the prestigious journal *PLoS ONE*, a team of researchers including the University of Iowa's Chris Jones—the guy who calculated Iowa's fecal-equivalent population—further demolished agribusiness claims that farm runoff is but one factor feeding the Gulf of Mexico algae blooms. Using data collected from streams throughout the Corn Belt, they calculated that between 1999 and 2016, Iowa alone contributed on average 29 percent of the nitrogen entering the entire Mississippi River—and as much as half in some years. Worse still, the state's five-year average contribution spiked by *40 percent* over the study period, suggesting that Iowa's nutrient-pollution problem is getting worse, not better.

The dead zone is a consequence of industrial agriculture's triumph over the prairie—as much a creation of the U.S. food system as the McDonald's dollar menu. As the environmental writer Richard Manning put it in 2004: the dead zone's routine annual appearance, blotting out life over a swath of the continental United States' most productive fishing region, means that a "high-quality source of low-cost protein is being sacrificed so that a source of low-quality, high-input subsidized protein can blanket the Upper Midwest."

Gulf of Mexico hypoxia was first documented by scientists in the 1970s, and the fossil record shows no signs of massive dead zones until the 1950s. Since this phenomenon is a relatively recent development, scientists are still assessing its long-term effects on the gulf ecosystem. Early evidence is alarming. Louisiana's celebrated brown shrimp fishery, worth about $30 million per year, is one well-documented case. "On the Louisiana shelf, hypoxia-induced habitat loss has decreased the range of brown shrimp by 25 percent since the early 1980s," the National Centers for Coastal Ocean Science reports. A 2017 study led by Duke University researchers found that exposure to oxygen-poor water slows the growth of brown shrimp and pushes them to congregate at the edge of the hypoxic zone, where fishing fleets catch them in earlier stages in their growth. The result: a shift to smaller, less valuable shrimp in years with big dead zones.

Another study, this one published in 2015 by George Mason University researchers, found that dumping huge amounts of nitrogen and phosphorus into the gulf generates ecosystem winners and losers. Some species, such as the highly prized red snapper, are harmed by increased hypoxia; others, such as jellyfish, proliferate. Megadoses of plant-feeding chemicals mean more plant growth and thus more food for species at the bottom of the food chain. Other research finds evidence that regular exposure to low-oxygen zones harms the reproductive capacity of certain fish species.

While the ecosystem impacts of the annual dead zone have yet to be fully understood, there's no denying that the Corn Belt way of farming is imposing huge and unpredictable changes on a vital food-producing ecosystem one thousand miles to its south. And climate change will very likely continue intensifying those changes. Of course, ever-fiercer spring storms—loaded with evaporated water from the gulf—are primed to keep pounding the Corn Belt's base soils, sending a vast load of fertilizer down the Mississippi and into the Gulf, ready to feed more algae. In addition, as water warms, it holds less oxygen, compounding the effect of oxygen-trapping algae blooms.

One of the ironies of both freshwater and coastal algae blooms is that their relationship with climate change runs both ways. A 2018 paper by a team that includes researchers from the EPA and the University of Minnesota found that algae-riddled lakes are much bigger emitters of the potent gas methane than was previously known. Methane is a greenhouse gas with about 30 times the heat-trapping effect of carbon dioxide. As for coastal hypoxic zones, they are significant emitters of nitrous oxide, a greenhouse gas with 296 times the heat-trapping power of carbon dioxide. These are examples of what researchers call a "positive climate-feedback loop"—they're accelerated by climate change, and they in turn help drive it.

As Cruse and I headed back to Ames on that early-summer soil-damage tour in 2019, I asked him about a technology fix. Couldn't, say, gene-edited crops engineered to shrug off drought bail us out?

He chuckled. "Formula One is a great technology, and when you put it on the Indianapolis Speedway, it does marvelous things. But on a dirt road behind the house, in a pasture? Same technology, but it just doesn't perform."

"You can't make something from nothing," he said. Food ultimately draws its nutrients from the soil, and degraded soils produce crops with lower levels of protein and other nutrients. The kinds of technology that can make a difference, he said, are practices that hold soil in place and build up its carbon content.

Failing that, I asked him, how long can we go on like we are? He thought for a second. "When you look at the averages of soil loss, they look bad, but it's not the averages that get you—it's the extremes," he said. Climate change promises to keep visiting more and ever wilder spring tempests upon the Corn Belt, more summer heat waves, more droughts.

He declined to predict how long the region's farmers could continue wringing bumper crops out of this precious but dwindling cache of soil. But its vulnerabilities are already showing, and the ripple effects are scary. He pointed to the course of the Arab Spring pro-democracy movements in the early 2010s. The Arab Spring was triggered in large part by food riots; the explosive growth of crop-based biofuels, Wall Street speculation, and poor harvests in several drought-plagued growing regions across the globe had caused food prices to spike, squeezing urban population across the Middle East. Then the Corn Belt drought of 2012 slammed U.S. corn yields; since the United States supplies nearly 40 percent of the corn that trades globally, the losses reverberated through global markets, providing another turn of the screw just as the early hopes of the Arab Spring movements were fading.

Cruse said he expected more spring rains to sacrifice more soil, making the land ever-more vulnerable to drought. "It's a snowball running downhill," he said. As we headed back to Ames, we gaped at several more ephemeral gullies, waiting to be refilled with some of the globe's best soil and planted with corn.

7

The Big Lift

My erosion tour with Rick Cruse on that hot day in June 2019 demonstrated to me that relying on vast corn and soybean fields, interrupted by the occasional cluster of indoor hog factories, is pushing the Corn Belt over an ecological cliff. That very same day, I also saw a different way forward in action.

After dropping off Cruse in Ames, I got back on the road to visit a farmer I kept hearing about named Tom Frantzen, who runs a diversified organic operation in Chickasaw County in northeast Iowa, a few hours' drive from Ames. Frantzen, Cruse told me, farms in a way that keeps soil in place. In a harried phone conversation to set up the interview, Frantzen let me know he was slammed: scrambling to take advantage of a rare dry spell to catch up on planting. But he was eager to show me his farm nonetheless, and promised to take me to an example of soil erosion a few miles from his farm that "needs to shock the country."

As I headed north and east on highways through soggy farmland, my now-trained eyes began to see erosion everywhere. After passing through the tiny town of New Hampton, near Frantzen's farm, the situation seemed to get more extreme: field after field, gently sloping toward the road, with gullies slicing through the young, struggling corn. The roadside ditches were ecological crime scenes, heaped with mud.

I pulled into a driveway off a dirt road to find Frantzen's modest farmhouse opposite several barns under the waning dusk light. The air smelled alive, sweet, slightly pungent—the scent of livestock kept at a scale that matched the surrounding landscape's ability to handle their waste. Unlike most farms in the twenty-first century Corn Belt, which specialize in crops or livestock, Frantzen's includes hogs and cattle along with around three hundred acres of field crops, grown mostly for feed. He markets the meat he produces through Organic Valley, a farmer-owned cooperative known mostly for dairy, but which also sells organic pork and beef.

Weary from a long day of driving, reporting, and immersion in the spectacle of ecological destruction, I hoped he would keep our interview short and invite me back in the morning. As I stepped out of the car, I heard the growl of a well-worn three-wheel ATV. Mustached, sixty-something, balding, and dressed in a rumpled button-down blue work shirt, Frantzen was beckoning me to jump aboard. Before I knew it, I was perched on the vehicle's side, white-knuckling the sidebar with one hand and holding a voice recorder in the other as we roared through the muddy barnyard. We lurched onto a trail beaten into a pasture. Above the motor's hum, Frantzen was saying something about rye. We moved over a slight rise in the land and on the other side I saw it: a field of rich green stretching to a line of trees hundreds of yards back. We stopped at the field's edge, and Frantzen jumped off the ATV, grabbed something from the back, and headed into the rye, which swayed lightly in the gentle wind, its thickly planted stalks reaching his chest.

Surrounded by his rye, he produced a sign naming the variety he had planted (Brasetto) and the small seed company that distributes it (Albert Lea Seed). Frantzen was ready for his photo op, and I complied, snapping a shot in the low light as the sunset gathered itself behind him.

Suddenly, I felt energized. In a landscape of beaten-up fields and sad corn crops, here was a lush, tall blanket of vegetation. And Frantzen, despite a frantic day in the field, was excited to make his pitch: that adding rye to the corn-soybean rotation is a powerful antidote to the ravages of squandered soil and fouled water that's eating away at the Corn Belt.

He started with a brief history of how Iowa farmers came to rely so heavily on corn and soybeans. When U.S. settlers broke ground in the 1860s, they tried winter wheat, which is planted in the fall and harvested in the summer. One advantage was that it kept the ground covered through those gully-washing spring storms. They enjoyed some good harvests, and the eastern United States, immersed in the Civil War, offered a robust market. Soon, however, wheat couldn't hold up to northern Iowa's brutally cold winters and hot, humid summers—weather that nurtured a habitat for insects and fungal diseases that, once established, "basically wiped out the winter wheat trade," Frantzen said. "Typically we're just too humid, too wet, and too cold."

Ever since, Frantzen continued, farmers in the region have fixated on what he calls "spring-seeded annuals," crops sown in the late spring and harvested in the early fall. For about the state's first century, those were corn and oats, which turned into corn and soybeans in the 1970s. All those crops, of course, leave the ground bare when the snow melts in the late winter, exposing them to weeks of potentially soil-destroying rains. Until the mid-twentieth century, farmers in the region did rotate in fall-planted hay crops (mainly clover and alfalfa) as cattle fodder, and that buffered around a third of their land from spring storms. But that practice withered away in the post–World War II years, when cows left the landscape for feedlots and soybeans took hay's place in their feed rations. What remained was a vast carpet of corn and soybeans taken away in the fall, prepping the ground for spring erosion events.

Frantzen motioned to his rye. Unlike those winter-wheat varieties of yore, he said, rye can handle northern Iowa's brutal winters. "I planted this on the twentieth of September, and by the twentieth of October, it was green," he said. Fall-planted crops go dormant over the winter, and then surge back to life when the soil thaws. "It protected the ground all winter, all spring, from torrential rains that did catastrophic erosion, unspeakable erosion—erosion that's a *national disgrace*," he thundered. "You saw that erosion today!"

Frantzen began riffing about how farmers could work winter rye into their rotations, devoting a third of their land to it every year, breaking up the corn-soybean duopoly. When it's grown in Iowa these days, rye typically serves as a cover crop: farmers plant it after the corn-soybean harvest and apply herbicides to kill it in the spring without harvesting the grains, before planting the next round of cash crops. Cover crops are an effective strategy for protecting the soil and building up organic matter, but as I saw in my sojourn through the countryside, they remain vanishingly rare. In the winter of 2017–18, they covered just 3 percent of Iowa's farm acres. Frantzen wanted to make sure I knew his rye was planted for *grain*—meaning that it provides the same soil-protection service as a cover crop, but also delivers a payday when it's harvested. That's vital, he said, because without such a cash incentive, "way too few" farmers were planting winter-hardy crops.

Frantzen farms just forty miles from the Minnesota border. In this northern region of the Corn Belt, it's tricky to grow a successful cover crop, because you have to wait until after the October-November corn-soybean harvest to plant it, he said. If winter comes early, the plants don't establish before frigid temperatures hit, which leads to paltry growth when spring comes—and money and time wasted. As a result, farmers in the northern Corn Belt tend to be wary of cover crops, and so they leave the ground bare.

Since it's planted in September, rye has plenty of time to develop enough to survive the harsh winter. Rye helps broaden the window for cover crops, too: once it's harvested in July, farmers have much more time to establish a cover crop ahead of winter, protecting the ground before it's planted with corn or soybeans in the fall. In other words, a single July planting of rye offers the potential to keep the soil buffered by vegetation for two straight years—the first year with the rye itself acting as a buffer; the second by extending the window for planting cover crops. A rye-corn-soybean rotation would protect as much as two thirds of the ground per year—a major step forward.

One of the things that has held back rye in the Corn Belt's hot, humid summers, Frantzen said, is that it's susceptible to a fungus called ergot, which reduces the quality of the grain, and therefore the price it fetches in the market. "There's reason to believe the ergot fungus caused the witchcraft hysteria in seventeenth-century Europe," he said, noting that Europeans of the era consumed loads of rye. "You take ergot and refine it, and you get LSD," he noted. (He quickly dismissed the notion that *that* could become a new income stream for Iowa farmers.) The giant seed companies like Bayer and Syngenta have focused their resources on breeding corn and soybeans adapted to Corn Belt soils and climates. Rye has emerged as a kind of orphan crop. But the small Minnesota seed house Albert Lea markets a hybrid rye that's bred in Germany (without transgenic technology) for ergot resistance and high grain yields.

While I was still digesting all this, Frantzen moved on to another advantage. "Let's pretend you are a nesting bird, like a pheasant," he said. "Well, let's look around the area and see where you're going to reproduce." He nodded to the distant right: "Well this hay field over there looks interesting, doesn't it?" Indeed, it did—it was a field of clover, at least waist-high. "I'm gonna go over it with a disk mower at fourteen miles per hour—by this time tomorrow, it will be six inches off the ground. If you have a nest in there, you're dead." He pointed to a distant field over on the other side of a fencerow behind the rye. It was a brown field faintly lined green. "What about my neighbor's corn and bean field? Not much there. How about the fencerow? Well, that's pretty good, but it's also pretty good hunting ground." He noted that nesting birds need wide areas of thick foliage and months of minimal disturbance. Frantzen gestured to the rye around him, noting that it had been a robust habitat since March and would not be harvested until July. "If you can't reproduce in that amount of time and get the hell out of Dodge, you're probably gonna go extinct," he said.

Frantzen's vision—large stands of summer-harvested winter rye woven through the Corn Belt, providing a habitat for birds—could be important. In a landmark 2019 study published in *Science*, researchers from Cornell and the USGS estimated that the U.S. bird population had plunged by nearly 30 percent since 1970. The paper cited widespread habit loss, including monocrop farming, and ubiquitous pesticide exposure as key factors driving the alarming decline. Specifically in the massive swath of grasslands that once flourished between Alberta, Canada, and central

Mexico—centered in the Corn Belt—bird populations plunged by 40 percent between 1966 and 2013. As Frantzen noted, to keep the great bulk of the Midwest essentially bare from the fall harvest to the early-summer establishment of corn and soybean crops gives birds no place to be. More winter rye, which requires little or no insecticides, could help.

Because of its quality and yield, ergot-resistant rye grown in the Corn Belt could be a major seller, used for whiskey, beer, and—the biggest potential market of all—hog and cattle feed, he said. As the light waned, Frantzen beckoned me back on the ATV, and we zoomed back to the barnyard, where he wanted to show me the feed trial he was running. We pulled up to a row of outdoor pens, each holding around fifteen hogs, with plenty of room for them to frolic. Typically, hogs get a mix of three parts corn to one part soybean meal in their rations. In several of the pens, Frantzen had replaced half the corn with rye stored from last year's harvest, as part of a feeding trial with the Practical Farmers of Iowa, a group that promotes sustainable agriculture practices. So far, he said, the rye-fed pigs were gaining weight at the same rate as the ones on the typical diet. The variety of hybrid rye Frantzen grows is booming in Europe, and is rapidly gaining ground as feed in the vast German and Danish hog industries.

Frantzen let himself into one of the pens. As nearly full-grown hogs jostled around him, grunting and snorting, he opened the lid on a feeder and plunged his hand in, producing a palmful of ground corn, soybeans, and rye—the crop mix that, if grown together carefully, he thinks could revive the Corn Belt's soil and prepare it for the cataclysms of the warming climate.

"I'm sixty-seven years old and *sick* of what the country looks like," Frantzen scowled, referring to the sea of eroded fields that surround his farm. "I think rye can change it."

In many ways, Tom Frantzen is a Corn Belt outcast. He farms organically in a landscape dominated by agrichemicals; he integrates livestock with crops while the vast majority of his peers typically choose one or the other;

and he adds a third crop to the corn-soybean rotation. But a growing weight of science demonstrates that his basic principle—adding diversity to a landscape sorely lacking in it—could solve the region's mounting soil- and water-quality crises.

Consider water quality a kind of leading indicator of a farming system's sustainability. When farms lose quality topsoil, they also lose their ability to capture water, leaving more to rampage across the landscape in the form of floods, which in turn pick up the chemicals that have been applied to bare soil. Those chemicals then turn up in streams, rivers, lakes, river deltas, wells, and municipal water plants, leading to cascading effects such as those I saw in Iowa's toxin-plagued Pine Lake.

As Frantzen noted that day on this farm, Iowa's fields have gone from native prairie to a brief attempt at winter wheat to the rise of what he called "spring-seeded annuals." But before the rise of industrial agriculture, Iowa maintained a fairly diverse agricultural landscape.

The ultimate spring-seeded crop on the American continent is, of course, corn, which was first domesticated in southern Mexico and Central America and had been widely planted in swaths of North America by the time of colonial settlement. Corn, it turned out, was well adapted to the region's climate, and is a good feed for pigs. For a long time, oats were a crucial part of the mix, in part because they were the preferred fodder for the draft horses that until the 1940s did the heavy work on Iowa's farms. Planted in the late spring and harvested in the fall, oats were often sown with clover, which would typically be left to grow after the oat harvest, providing winter cover and spring hay. And farmers would often devote fields to another hay crop, alfalfa, planted in the fall and fed to cattle and dairy cows.

So, most farmers in pre–World War II Iowa—a microcosm of the greater Corn Belt—employed a system involving two livestock species (cows and pigs), grown on pastures, interspersed with three main crops grown in rotation (corn, oats, and hay). Rotating the crops interrupted weed and insect-pest patterns, providing a handy advantage, given that today's herbicides and pesticides weren't readily available, if they existed at all. And those hay crops, alfalfa and clover, are legumes, so they trap nitrogen from the air and leave some behind in soil—an important function in the age before industrial fertilizers. More important, they kept a significant portion of the ground covered over the winter, protecting it from harsh storms and limiting erosion, as did the millions of acres devoted to pasture for hogs and cattle.

With the end of the war, everything changed. The rise of tractors meant no more need for horses, so oats began to vanish from the landscape. By 1960, a previously marginal high-protein feed crop called soybeans began to take their place. Soybeans are much higher in protein, and easier to store and transport, than alfalfa and clover, so they displaced hay in the rotation. By the 1970s, the Corn Belt had mostly completed its evolution into the corn, soybean, and hog factory it is today. A system of multiple crops grown in conjunction with two livestock species had dramatically simplified.

There is considerable evidence that this great simplification has triggered the decline in Iowa's water quality—that is, that pre–World War II land-use patterns led to much cleaner water leaving farms. Analyzing a 1955 document on water quality from the Iowa Geological Survey, which tracked nitrate concentrations measured in Iowa waterways in the first half of the twentieth century, the University of Iowa's Chris Jones found "roughly a 10-fold increase from a century ago and a 3- to 4-fold increase since 1950," compared with late 2010s levels. And looking at nitrates in streams and rivers in terms of parts per million greatly undersells the gross amount of the chemical that's escaping farms, Jones added. That's because there's more runoff water than in the 1950s (the soil is less permeable and there is more erosion, so the soil is less adept at trapping water), and all the runoff carries more nitrate. That is, not only does the water carry much higher concentrations of nitrate; there is also as much as three times more water flowing from farms into streams—the result of more frequent and powerful storms and more advanced tile drainage.

Another way to gauge the quality of the water leaving the Corn Belt's farms is to look south to the Gulf of Mexico, where the runoff ultimately gathers. As we saw in chapter 6, the area's dead zones developed roughly simultaneously with the winnowing down of crop diversity and the rise of industrial hog farms in the Corn Belt.

As Jones notes, one obvious driver is the rise in fertilizer use. New, higher-yielding varieties of hybrid corn needed more nitrogen to reach their maximum output potential. Meanwhile, soybeans, like all legumes, trap nitrogen from the air and deposit it in the soil in a form that plants can use, but they do so at a rate much lower than that of the hay crops they displaced—clover and alfalfa. With less nitrogen fixed by legumes, farmers had to apply more to their corn. Jones shows that in 1960, farmers were applying nitrogen to their corn at an average rate of about 20 pounds per acre; by 1980, the average rate had jumped to 160 pounds, a level that has held and often been exceeded since. Rates for phosphorus, the fertilizer component that drives toxic algae blooms in freshwater bodies like lakes, doubled over that period.

It's true that while total fertilizer use exploded in the postwar years, corn yields did, too. Rising corn yields in the post–World War II era roughly track nitrogen fertilization rates. This relationship suggests that to restrict fertilizer use—and thus deposit less nitrate in water—also means restricting crop yields. That is, the only way to protect our water is to grow less food. This calculation serves the interests of agribusiness. Essentially, it presents polluted water in the Midwest as the necessary cost for a robust food supply. The agribusiness-allied Iowa Farm Bureau makes this case. When it's not denying the water-pollution crisis outright, the group stresses that "farmers must reach optimal crop yields and profitability to remain economically viable."

Even setting aside that farmers for most of the past forty years have greatly overproduced corn and soybeans, the argument that cleaning up water and saving soil means a major hit to food production is wrong. It turns out that high modern crop yields can be maintained and fertilizer use slashed by restoring biodiversity. Iowa State University agronomist Matt Liebman and a team of researchers have been testing this idea since 2002, when they put in test plots at the university's Marsden Farm in Boone County, just outside Ames.

The experiment is simple. In one portion of the farm, corn rotates with soybeans every year, replicating the pattern that dominates the region. In another, the researchers add a third crop to the annual mix: either oats or triticale, a wheat-rye hybrid, sewn with red clover, which is allowed to grow through the winter and harvested in the spring as livestock fodder. In the third and final set of plots, a fourth crop, alfalfa—also used for feed—joins the rotation. In all three, fertilizers and herbicides are used as needed to maintain robust crops.

The results, first summarized in a 2012 paper published in *PLoS ONE*, are stunning. The three- and four-year rotations required just 10 to 20 percent of the fertilizer and herbicide—and produced per-acre corn and soybean yields *greater* than those in the two-year rotation. As for runoff, the ecotoxicity of the water from the longer rotations—a measure of herbicide residues—measured at 0.5 percent of the levels of the two-year fields. Soil carbon levels, meanwhile, increased in the diverse rotations, and soil erosion rates plunged by 60 percent. While it's true that the longer rotations generated less corn and soybeans overall—because oats and alfalfa took their place some years—in terms of total crop output, the three- and four-year rotations produced 8 percent more than the corn-soy duopoly.

The kicker: the project's results suggest that the region's farmers won't take an economic hit from moving beyond growing just corn and beans. In short, growing a wider variety of crops requires more labor and management, but those expenses are balanced out by drastically reduced expenditures on agrichemicals.

So why don't farms switch en masse to these broader rotations? Natalie Hunt, a University of Minnesota researcher and a collaborator on the project, told me that the economic analysis assumed that the oats and alfalfa generated by the biodiverse plots would find a profitable use by being fed to cattle and hogs "on-farm or on neighboring farms." A farm that planted an alfalfa rotation, for example, could "harvest" it by simply turning cattle loose on it for munching; which would enhance the income stream from beef. That setup works best for diversified operations like Frantzen's, which are few and far between. Without nearby markets for alfalfa and oats, farmers can't simply diversify their crops; they also have to introduce cattle and/or pigs. That's a heavy lift.

Another obstacle, Hunt told me, is the "heavily taxpayer-subsidized crop insurance program that keep farmers locked into a corn- and soybean-producing system year after year, even when market prices are poor," as they have been for the past several years. She added, though, that if consumers demanded food from the Midwest that didn't pollute water and damage soil, the "market would respond pretty quickly." That is, if farmers could get a premium price for crops, meat, and milk "grown with biodiversity" or some such label, farmers would have an incentive to add them to their rotations.

Indeed, a small but growing cadre of farmers across the Corn Belt are finding ways to use biodiversity to enhance their bottom lines while also improving their soil. The way they deploy it depends on the particular ecological circumstances and challenges of their region. In June 2013, I visited one of the innovators: David Brandt, who farms 1,200 acres in the village of Carroll, Ohio, population 524, about a half-hour drive southeast of Columbus. Brandt operates at the Corn Belt's eastern edge, amid a vast expanse of corn and soybean fields broken up only by the exurban sprawl creeping in from Columbus.

Summer 2013 was an instructive time to visit. The previous growing season, farmers endured a historic drought that sliced crop yields throughout the Corn Belt. Statewide, production in Ohio's corn fields plunged by 22 percent beneath normal levels that season. In spring 2013, weather lurched from too dry to too rainy: a series of heavy storms delayed planting and caused erosion and water pollution, in an increasingly familiar pattern. Such drought-flood transitions are a kind of stress test for farmers dealing with a warming climate and ever-wilder weather, and I was happy for the chance to see how Brandt had fared.

Chatting with him outside his barn on a sunny early-summer morning, he seemed almost like a farmer in a movie—what a casting director might call "too on the nose." He's a beefy man in bib overalls, a plaid shirt, and well-worn boots, with short, gray-streaked hair peeking out from a trucker hat over a round, unlined face ruddy from the sun. On first meeting him, I noticed that he deployed a Midwest monotone to give short, clipped answers to my questions. Until I brought up cover crops.

In Frantzen's corner of the Corn Belt, cover crops are tricky. But Brandt farms about one hundred miles to the south, where there's more time to establish a good stand before winter settles in. And whereas hybrid rye gets Frantzen excited, it was the topic of cover crops in general that got the laconic Brandt talking. Driving in, I could not tell Brandt's fields from those of his neighbors; it all looked just about identical to the carpet of corn and soybeans that stretches from eastern Ohio to western Nebraska. But three months before, Brandt explained, his land would have stood out. While the neighbors' fields lay bare, Brandt's teemed with a mix of as many as fourteen different plant species—what Rafiq Islam, an Ohio State University agronomist who works closely with Brandt, calls a "cover crop cocktail mix."

Like Frantzen and those researchers at Iowa State, Brandt adds a third crop to the corn-soybean mix—in his case, wheat. But it's what he calls "my covers" that drives his system. Radishes, he explained, form a thick, cylindrical root, about the size of an adult's forearm, that plunges into the soil. As the root decays, it not only feeds the teeming population of microorganisms that make up the soil biome; it also leaves behind a big air pocket that breaks up the dirt and makes pathways for water to percolate downward. Legumes like hairy vetch grab nitrogen from the air and deposit it in nodules in their roots, adding fertility. Grasses including rye grow prolifically, sucking carbon from the air and into their roots and stems: more food for underground critters. (Rye serves Frantzen as a promising cash crop; for Brandt, it's purely soil food.)

"Our cover crops work together like a community. You have several people helping instead of one, and if one slows down, the others kind of pick it up," he explained. "We're trying to mimic Mother Nature." Cover crops have helped Brandt slash his use of fertilizers and herbicides. That season, he told me, half his corn and soy crop was flourishing without any of either; the other half had gotten much lower applications of those pricey additives than what crop consultants in his county recommend.

Cover crops are a linchpin of organic farming, but Brandt's not trying to go organic. He prefers the flexibility of being able to use conventional inputs in a pinch. He refuses, however, to compromise on one thing: tilling. Brandt never, ever tills his soil. Ripping the soil up with steel blades creates a nice, clean, weed-free bed for seeds, but it also disturbs soil microbiota and leaves bare ground vulnerable to erosion. No-till farming, as it's known, became much more common with the rise of corn and soybean varieties genetically engineered for herbicide resistance, which has allowed farmers to use chemicals instead of the plow to control weeds. The seed-agrichemical industry hypes the rise of herbicide-driven no-till as an environmental triumph. But the brand of no-till most prevalent in the Corn Belt is periodic—farmers typically forgo tilling before planting soybeans but revert to the plow before planting corn. Every time they plow, most of the long-term soil-building benefits (including carbon sequestration) are vaporized.

Standing in the shade of Brandt's barn that June morning, I heard a commotion in the nearby warehouse where he stores his cover-crop seeds. Turns out that I wasn't the only one visiting Brandt's farm. When I had scheduled my trip, he had forgotten to tell me that the Natural Resources Conservation Service (NRCS)—a branch of the U.S. Department of Agriculture that grew out of Dust Bowl–era efforts to preserve soil—was holding a training for its agents on how to talk to farmers about cover crops and their relationship to soil.

He invited me to follow him into a warehouse, where fifty-pound bags of cover-crop seeds lined one wall and three dozen NRCS managers and agents, from as far away as Maine and Hawaii, had gathered along tables facing a projection screen. Brandt, it turned out, was scheduled to make a speech that morning. He took his place in front of the crowd, I took mine in it. Presenting slides of fields flush with a dizzying combination of cover crops, he delivered a spirited harangue while we listened raptly and nodded approvingly. It began to feel like a revival meeting.

"We want diversity," Brandt thundered. "We want colonization!"—that is, for plants to take root so densely that little or no ground remains exposed. While the cash crop brings in money and feeds people and livestock, he told the agents, the off-season cover crops feed the soil and the hidden universe of microbes within it, accomplishing much of the work done by chemicals on conventional farms. And the more diverse the mix of cover crops, he declared, the better the whole system works. Brandt pointed to the heavy, mechanically operated door at the back of the warehouse and then motioned to the crowd. "If we decide to lift that big door out there all together, we could do it," he insisted. "If I try by myself, it's going to smash me."

Soon, we all filed outside and walked past the Brandt family's four-acre homestead garden. Chickens moved about in clumps, pecking the ground, *bawk-bawk-bawking* and getting underfoot. In an open barn nearby, a few cows munched hay. I saw pigs rooting around in another open barn thirty or so yards away and started to wonder if I hadn't stumbled into a time warp, to the place where they shot the farm scenes in *The Wizard of Oz*. As if to confirm it, a cow emitted a plaintive moo. Brandt's livestock are a hobby—"freezer meat" for his family and neighbors. As we peered around the barns, we saw the edges of his real operation: a pastiche of fields stretching to the horizon.

Before we could get our hands in the dirt, Brandt wanted to show us his farm equipment: the rolling contraption he drags behind his tractor to kill cover crops ahead of the spring and the shiny, fire-engine-red device he uses to drill corn and soy seeds through the dead cover crops directly into the soil. As some NRCS gearheads peppered him with questions about the tools, he beamed with pride.

He explained his system. After the fall harvest, he plants his entire farm in cover crops. They go dormant when winter sets in and then burst back to life after the spring thaw. In May, a week or so before it's time to plant the season's corn, soybeans, and wheat, it's time to kill the cover crops. When conditions are right—not too wet—he attaches a special implement called a crimper to his tractor, which kills them and creates a mat of plant matter over the land. In wet years, he uses herbicides to do the deed—the only use for chemical weed-killers on his farm. He then sows his cash-crop seeds directly into the untilled soil, using another implement that drills the seeds through the aboveground mat. The seeds germinate, while the decaying plant matter prevents weeds from establishing. Thus the cover crops not only feed the soil; they also provide all the weed control he needs.

Soon it was time to see his system in action. We all filed onto an old bus for a drive around the fields, with Brandt at the wheel. An ag nerd among professional soil geeks, I felt like I was back in elementary school on the coolest field trip ever. An almost giddy mood pervaded the vehicle as Brandt steered us to the side of a rural road that divides two cornfields: one his, and one his neighbor's.

We started in Brandt's field, where we encountered waist-high, deep-green corn plants basking in the afternoon heat. The night before, he said, there had been a heavy rainstorm. A mat of withered leaves and stalks covered the soil, remnants of the winter cover crops that had, as advertised, kept the field devoid of weeds. At Brandt's urging, we scoured the ground for what he called "haystacks," little clusters of dead, strawlike plant residue bunched up by earthworms. Sure enough, the stacks were everywhere. Brandt scooped one up, along with a fistful of black, crumbly dirt, still moist from a recent rain. "Look there, and there," he said, pointing into the dirt at pinkie-size wriggling earthworms. "And there go some babies," he added, indicating a few so tiny they could curl up on your fingernail.

Then he directed our gaze to the ground where he had just scooped the sample. He pointed out a pencil-size hole going deep into the soil: a kind of worm thruway that invites water to stream down. I doubt I was the only one gaping in awe, thinking of the thousands of miniature haystacks around me, each with its cadre of worms and its hole into the earth. I looked around to find several NRCS people holding their own little clump of live dirt, oohing and aahing at the sight.

Holding a handful of loamy soil, Brandt explained that he habitually tests his dirt for organic matter. When he began renting this particular field two seasons before, its organic content stood at 0.25 percent—a pathetic reading in an area where, even in fields farmed conventionally, the level typically hovers between 1 and 2 percent. In just two years of intensive cover cropping, this field has risen to 1.25 percent. Within ten years of his management style, he adds, his fields typically reach as high as 4 percent, and with more time can exceed 7 percent.

Then we crossed the street to the neighbor's field. Here, the corn plants looked similar to Brandt's, if a little more scraggly and a slightly paler shade of green. The soil, though, couldn't have been more different. The ground, unmarked by those haystacks and mostly bare of plant residue altogether, was seized up into a moist, muddy crust. When I scratched the surface, I found the dirt just below was dry. Brandt pointed to a pattern of ruts in the ground, cut by rainwater that had failed to absorb and gushed away: the same effect that gives rise to ephemeral gullies when the ground is completely uncovered during spring storms.

Brandt's land had managed to trap the previous night's rain, giving the corn a reservoir to draw on until the next storm. His neighbor's plot had lost not just the precious water but the untold chemical inputs that it carried away. (Land in this portion of Ohio drains into the Ohio River, ultimately making its way to the Mississippi River and the Gulf of Mexico.)

After the field trip, we reconvened in the warehouse for some demos. At a table at the front of the room, an NRCS man dressed in country casual—faded jeans, striped polo shirt, baseball cap—dropped five clumps of soil into water-filled beakers: three from farms managed like Brandt's, with cover crops and no tillage; the others from conventional operations. The Brandt-style samples held together, barely discoloring the water. The fourth one held together too, but for a different reason: unlike the no-till, cover-crop samples, which the water had penetrated, this one was so compacted from tillage that no water could get in at all. The fifth one disintegrated before our eyes, turning the water into a cloudy mess that the NRCS presenter compared to "last night's beer." Both the fourth and fifth samples were evidence of carbon-poor soils common in the Corn Belt.

Another demo showed how water percolated through Brandt's gold-standard dirt as if through a sieve, picking up little color. In the conventional soil, it pooled on top in an opaque mess, demonstrating that the soil's density, or compaction, can cause runoff.

The tour of Brandt's farm impressed me. Later, I checked in with some scientists familiar with the operation to make sure what I saw wasn't too good to be true. The eminent soil scientist Rattan Lal of Ohio State University told me that he had been observing Brandt's farm for years, and that his combination of continuous no-till and cover crops had proved able to rapidly store carbon in soil, just as Brandt claimed. Lal, a member of the International Panel on Climate Change, is so impressed with Brandt's methods that he regularly brings farmers from other countries to tour the farm, he told me. If all U.S. farms adopted Brandt-style agriculture, Lal estimated, they could suck down as much as twenty times more carbon than they currently do—equivalent to taking nearly 10 percent of the U.S. car fleet off the road.

That represents a real potential to contribute to slowing down climate change, though at a rather limited scale. But more immediately, Brandt's style demonstrates massive promise for helping farmers *adapt* to climate change: carbon rich soils stay put during storms and hold water during drought. Lal's colleague Rafiq Isla, who has also been studying Brandt's farm for years, said that it regularly achieves crop yields that exceed the county average, and during the brutal 2012 drought, Brandt's fields stayed strong at about 90 percent of their long-term average output, while nearby farmers saw yields drop by 50 percent or lost their crop entirely.

As the field day in 2013 ended, I imagined the NRCS reps fanning out across the Midwest to bring the good news about cover cropping and continuous no-till. And I wondered: Why aren't these ways spreading like prairie fire, turning farmers into producers of not just crops but also rich, carbon-trapping soil resilient to floods and drought?

I put the question to Brandt. His own neighbors weren't exactly rushing out to sell their tillers or invest in seeds, he said. They tended to see him not as a beacon but rather as an "odd individual in the area," he said, his level voice betraying a hint of irritation. Sure, his yields are impressive, but federal crop payouts and subsidized crop insurance buffer his neighbors' losses, giving them little short-term incentive to change, he said. (For his part, Brandt refuses to carry crop insurance, saying it compels farmers "not to make good management decisions.") Plus the usual way is easier. Using diverse cover crops to control weeds and maintain fertility requires much more management, and more person-hours, than relying on chemicals.

Longer term, though, Brandt did express hope. Over the next twenty years, he envisioned a "large movement of producers" adopting cover crops and no-till farming in response to rising energy costs, which could make fertilizer and pesticides (synthesized from petroleum and natural gas), as well as tractor fuel, prohibitively expensive.

Mark Scarpitti, an NRCS agronomist in attendance, agreed. He acknowledged that in Brandt's corner of Ohio, the old saw that the "prophet isn't recognized in his own hometown" largely holds, though a "handful" of farmers were catching on. Nationwide, he added, "word's getting out" as farmers like Brandt slowly show their neighbors that biodiversity, not chemicals, is their best strategy.

Sure enough, during the NRCS meeting, another local farmer stopped by to pick up some cover-crop seeds. Keith Dennis, who farms around 1,500 acres of corn and soy in Brandt's county, and who started using cover crops in 2011, says there are quite a few folks in the county watching what Brandt is doing, "some of them picking up on it." Dennis has known about Brandt's work with cover crops since he started in the 1970s. I had to ask: If he saw Brandt's techniques working then, what took him so long to follow suit? "I had blinders on," he replied, adding that he saw no reason to plant anything but corn and soybeans. "Now I'm able to see that my soil had been suffering severe compaction," he said. "Because it wasn't alive."

8

The Future of the Farm

In the winter of 2018–19, California farmers hit the jackpot. An onslaught of more than thirty atmospheric rivers, none of them large enough to trigger significant floods, delivered a massive snowpack that measured more than 60 percent above average. "With full reservoirs and a dense snowpack, this year is practically a California water-supply dream," Karla Nemeth, director of the state's Department of Water Resources, declared at an April ceremony.

By September, farmers in the San Joaquin Valley were living a water-supply nightmare. CALIFORNIA FARMERS FACE "CATASTROPHIC" WATER RESTRICTIONS. CAN THEY ADAPT TO SURVIVE? declared a headline in the *Sacramento Bee*. Throughout the valley, especially in the Tulare region, water was in short supply. Farmers were experiencing their last season of unrestricted groundwater pumping, and staring into the abyss of California's Sustainable Groundwater Management Act (SGMA), the 2014 law requiring that overpumped basins like the Tulare restrict pumping starting in 2020, so that aquifers come into balance by 2040.

The adjustment will mean draconian cuts in irrigation, and "that will mean a lot fewer pistachios, grapes, almonds and tomatoes," the *Bee* reported. Water anxiety was sweeping the valley, the *Bee* continued:

> Eric Limas, who runs a groundwater agency in the Pixley area of Tulare County, says his water allotment will be downright frightening: Farmers on his turf will have to curtail their groundwater usage by 40 percent eventually. "You're talking devastation here, in the catastrophe spectrum," Limas said.

In other words, even in increasingly rare "good" snowpack years—even in "dream" ones, like 2018–19—farmers in the nation's most productive agricultural valley rely on pumped groundwater to irrigate their crops. That means hard choices are ahead. "You're farming almonds and I'm farming carrots. Your ability and willingness to pay for water is greater than mine. That's the economics," one Kern County water manager told the *Bee*. Translation: nut crops, with their massive embedded investments and booming overseas markets, will grab the most water, at the expense of vegetables.

The plight of water-starved San Joaquin Valley farmers barely penetrated the national conversation at the time. But California's mounting water scarcity will be impossible to ignore when drought recurs, as it inevitably will. Meanwhile, as water stress continues to force a shift from annual crops like vegetables to perennial ones like almonds in the San Joaquin Valley, it might be tempting to hope that other regions in California will fill any shortfall in the supermarket produce aisle.

For the full picture of the state's coming agricultural crisis, we need to consider two other California regions that are significant suppliers to the national food market: the Salinas Valley, known as the "Salad Bowl of the World"; and the Imperial Valley, which specializes in fresh winter produce. They, too, face severe and intensifying water jeopardy.

A rifle-shaped slice of land between two mountain ranges just south of Monterey Bay, off the state's central coast, the Salinas Valley churns out more than half the salad greens and half the broccoli consumed in the United States. Its leafy-green dominance has earned it its nickname. But unlike the Central Valley, it suffered almost no consequences from the 2012–16 drought. That's because farmers in the semi-arid Salinas, unlike their Central Valley counterparts, have no access to snowmelt diverted from a mountain range. Instead, they rely nearly 100 percent on underground aquifers, drought or no. And that means, in the short term, the state's vicious periodic snowmelt droughts can actually be economically beneficial: they don't affect production in the aquifer-irrigated Salinas, but they force water-scarce farms in the Central Valley to cut back on nonpermanent crops like lettuce, reducing supply and pushing up prices.

But all isn't well under those sunbaked fields teeming with ripe vegetables. For one thing, decades of heavy nitrogen-fertilizer use have left underground water widely contaminated with high levels of nitrate, placing a heavy burden on farmworker communities that rely largely on wells and lack access to filtration infrastructure. The USGS shows that 20 percent of the region's wells have nitrate levels over the legal limit. High housing prices, a spillover from the nearby Bay Area, apply another squeeze. The region's ninety thousand farmworkers live at an average residential housing rate of seven people per one-room apartment. Nearly 60 percent of the region's largely farmworker population live near or below the poverty line.

Worse, the region's aquifers, the lifeblood of its $8.2 billion ag economy and sole source of the region's drinking and irrigation water, are in a "state of long-term overdraft," a 2014 assessment from California's Water Foundation found. The paper notes that all eight of the Salinas Valley's aquifers were listed as among the most-stressed aquifers by the California Department of Water Resources. As irrigation wells draw down the valley's underground reservoirs, seawater from Monterey Bay seeps into the void, threatening to make the water too salty for farming. In response, in 2017, the Monterey County Water Resources Agency's board of supervisors imposed a moratorium on digging new wells in one large swath of the region, which enraged agricultural interests and will limit their future water access if the prohibitions remains in place. While efforts are afoot to halt seawater intrusion and bring the aquifers into balance, the process—if it succeeds—will be expensive and likely result in farmland needing to be fallowed or abandoned. This, in turn, could mean pricier salad greens and strawberries for the produce aisle.

Then there's the Imperial Valley, a bone-dry chunk of the Sonoran Desert bordering Arizona and Mexico in California's southeast corner. The

Imperial Valley (along with neighboring Yuma County in Arizona) churns out more than half the vegetables eaten by Americans during the winter. Major crops include broccoli, carrots, cauliflower, and, most famously, lettuce and salad mix. Overall, Imperial producers deliver about 14 percent of all leafy greens consumed domestically.

In terms of native aquatic resources, the Imperial Valley makes the Central Valley look like *Waterworld*. At least the Central Valley is bounded by a mountain range to the east that, in good years, delivers abundant snowmelt for irrigation. The Imperial sits in the middle of a blazing-hot desert, with no water-trapping mountains anywhere nearby. It receives a whopping three inches of precipitation per year on average; even the more arid parts of the Central Valley get five inches.

The sole source of water in the Imperial Valley is the Colorado River, which originates hundreds of miles northeast, in the snowy peaks of the Rocky Mountains. As it winds from its source down to Mexico, the Colorado delivers—in theory—a total of 16.5 million acre-feet of water to the farmers and 40 million consumers in seven U.S. states and northern Mexico who rely on it. (An acre-foot is the amount it takes to flood an acre of land with twelve inches of water—about 326,000 gallons.) The river has been oversubscribed for nearly a century; the 16.5 million acre-feet estimate derives from average flows recorded in the early twentieth century, before the river had become fully exploited by the western states. Since 1922, the average flow has been 14.8 million acre-feet. As of 2015, the ten-year average stood at 13.4 million acre-feet. That's because the Rocky Mountain snowpack that feeds the Colorado River declined by 41 percent between 1982 and 2017. It is expected to dwindle by another 30 percent by mid-century.

As the river is depleted, the farms in California's Imperial Valley retain a tight grip over 2.6 million acre-feet of water—more than half the state's total allotment and more than any other state draws from the river besides Colorado. Because it owns senior water rights based on a 1931 pact, the Imperial gets its allotments during low-flow years even when other regions see reductions. Whereas Central Valley farmers, reliant on vanishing snowmelt from the Sierras, saw their irrigation allotments slashed to zero during the Great Drought of 2011–17, growers in the Imperial Valley didn't miss a beat.

Pressure to share some of this water with the Southwest's burgeoning metropolises has been growing for decades. Under heavy political tension, the Imperial Irrigation District agreed to sell 200,000 acre-feet annually of water to fast-expanding San Diego in a 2003 deal that lasts until 2047. Calls to send Colorado River water to cities will mount as the climate warms and the Rocky Mountain snowpack recedes.

These tussles over control of the Colorado's flow become devastatingly visible at the shores of a beleaguered inland body of water that sits uneasily at the Imperial's northern edge: the Salton Sea, the original sin of the region's booming farming sector. Before the big irrigation projects that made the valley bloom, it was known as the Salton Sink—a depression sitting 235 feet beneath sea level. For millennia, it periodically captured floodwater from the then mighty Colorado, in the odd year when the river careened off its normal path to drain into the Gulf of California. These intermittent flows fed the area's ground with river silt, creating a fertile soil base.

In the early twentieth century, U.S. settlers began diverting a portion of the Colorado River west to the Salton Sink. In 1905, their designs worked all too well. A heavy snowpack in the Rockies combined with an engineering blunder to divert the river's entire flow into the Salton depression, an unintended deluge that lasted eighteen months. The modern Salton Sea was born: it instantly became and remains the state's largest lake. For years, it flourished. The valley's farmers claimed their massive share of the river's annual flow, now diverted in orderly fashion (after the rise of publicly funded, professionally managed aqueducts); drainage from farm irrigation would flow into the lake, more or less balancing evaporation and allowing the lake to hold its size. The Salton Sea emerged as a swank resort; pop stars from Frank Sinatra to the Pointer Sisters to the Beach Boys entertained there. The water teemed with fish, and it became a stopping point for migrating birds along the Pacific Flyway now that the vast, ancient network of rivers, lakes, and wetlands to the north in the modern Central Valley had been dammed and drained for farming.

Over time, moderate levels of salt from irrigation runoff built up in the lake, intensified by evaporation under the desert sun (which carries away water but leaves minerals behind). By the 1980s, the lake's salinity became too high to support most fish species, leading to massive die-offs. The resort withered as the lake began to emit the odor of rotting organic matter.

More recently, the San Diego water transfer forced the valley's farmers to use water more efficiently. The resulting widespread switch to drip irrigation meant less farm runoff, and thus less water flow into the Salton, which began to shrink in size (and grow even saltier) as evaporation outpaced inflow. As the Salton contracts, the desert sun quickly dries the exposed lake bottom, and the wind picks up the fine-particulate dust. Consequently, children in Imperial County endure asthma hospitalizations at twice the state's average rate. And as millions of fish corpses decay in the water, the lake periodically releases vast plumes of hydrogen sulfide, which can foul the air with a rotten-egg stench as far away as Los Angeles. If present trends hold, the Salton will continue shrinking, subjecting the Imperial Valley's residents—mostly farmworkers—to ever-more toxic air. State policymakers have been puzzling for decades over the best way to mitigate this slow-build ecological disaster; so far, the only clear consensus is that doing it right will require a "staggering" amount of Colorado River water.

The added burden of replenishing the Salton Sea looks especially onerous in light of the fact that the Colorado's flow has already proved inadequate to supply the broader region's needs. In a 2014 paper, U.C. Irvine and NASA researchers found that farmers and municipalities are quietly supplementing their river allocations by tapping wells and drawing water from underground aquifers at a much faster rate than had been known. Between December 2004 and November 2013, the Colorado Basin lost almost 53 million acre-feet of underground water, the equivalent of more than four annual flows of the Colorado River. That's an enormous fossil resource siphoned away in less than a decade.

"Quite honestly, we are alarmed and concerned about the implications of our findings," study coauthor Jay Famiglietti wrote at the time. "From a group that studies groundwater depletion in the hottest of the hot spots of water stress around the world—in India, the Middle East, and in

California's Central Valley—that says something." In short, Famiglietti writes, "The American West is running out of water."

Looked at from a global perspective, the calamity brewing in California is far from unprecedented. The United States is hardly the first modern nation to throw its lot with groundwater-driven desert agriculture for its sustenance. In the early 1980s, Saudi Arabia embarked on a bold project: it began to transform large swaths of desert landscape into wheat farms. For all their challenges, deserts also offer ample sunlight and cool nights, and harbor few crop-chomping insects, fungal diseases, or weed species. As long as you can strategically add water and fertilizer, you can generate bin-busting crops. And that's exactly what Saudi Arabia did. The oil-producing behemoth grew so much wheat for about two decades that it could take care of its own domestic needs while also supplying Kuwait, the United Arab Emirates, Qatar, Bahrain, Oman, and Yemen.

But starting in the mid-2000s, Saudi wheat production began to taper off. Soon after, it plunged. By 2015, the country had stopped producing wheat altogether, and instead opted to rely on global markets for the staple grain. What happened? To irrigate its wheat-growing binge, the nation tapped aquifers that "haven't been filled since the last Ice Age," *Bloomberg News* reported. And in doing so, it essentially drained them dry in the span of two decades.

The first sign of Saudi agriculture's water crisis began in the early 2000s, when long-established desert springs—ones that had "bubbled up for thousands of years from a massive aquifer system that lay underneath Saudi Arabia"—began to dry up. It had been "one of the world's largest underground systems, holding as much groundwater as Lake Erie," journalist Nathan Halverson reported in 2015. The town of Tayma, built on an ancient desert oasis, has been continuously inhabited for nearly three millennia. By 2011, its wells were "bone-dry," Halverson said—"drained in one generation."

In the meantime, farmers' wells, too, began to go dry, and they had to drill them ever deeper to keep the water flowing. By 2012, fully four fifths of the ancient aquifer had vanished. The Saudi government began to reconsider its make-the-desert-bloom ambitions, which have now turned to dust. (In the late 2010s, after the Saudi government cracked down on domestic agricultural water use, it began importing substantial amounts of alfalfa, a dairy cow feed, from, of all places, the Imperial Valley.)

Here in the United States, we've followed a similar strategy for fruit, vegetable, and nut production. We concentrate it in arid regions of California and irrigate it by diverting river water, sometimes over great distances, and by tapping massive aquifers, as the Saudis did. But climate change means less snow to feed rivers—and thus, to water farms—and more reliance on those overtapped underwater reserves. As we saw in chapter 1 of this book, it's impossible to know when the aquifers in California's vast Central Valley will go dry, because no one has invested in the research required to gauge just how much water is left. But the trend is clear. In large swaths of the region, the land is sinking at rates of up to eleven inches per year as underground water vanishes.

Unlike their Saudi peers, U.S. policymakers don't have the luxury of waiting until California's water runs out and then simply shifting to a reliance on imports. For one thing, our population is more than ten times larger. But another complicating factor lies in the kinds of crops that severe aquifer depletion will require us to replace. For the Saudis, it was wheat, which is grown in large quantities in the world's generally water-rich temperate zones. The fruits and vegetables the United States would need, though, grow most cheaply and abundantly in hot, dry zones, as long as plentiful water can be accessed.

The problem is that, like California, these regions tend to have their own severe water issues. Take Mexico, the source of 44 percent of U.S. fruit and vegetable imports. Two of its three main produce-exporting regions, Baja California and the Mexicali–San Luis Valley, sit directly south of California and face even more extreme problems. Receiving on average just three inches of rain annually, Baja farmers rely almost exclusively on groundwater for irrigation to supply the booming U.S. markets for strawberries, tomatoes, and berries. The region's four main aquifers are all severely overtapped, so much so that three of them have incurred high levels of seawater intrusion—the same problem that haunts the Salinas Valley, but at an even more advanced stage. Already, the Mexican government is having to subsidize desalination plants to make the pumped groundwater suitable for irrigation. Meanwhile, households in the four watersheds lack running water at rates ranging from 12.9 and 19.5 percent. The Mexicali–San Luis Valley, directly to the east of Baja, generates fruits and vegetables in the desert directly south of the Imperial Valley and also relies on the Colorado River for irrigation, but its water rights are much less privileged than the Imperial Valley's. In a recent survey of Mexico's water balance by World Resources Institute, those two regions ranked among Mexico's most water stressed, as did Sinaloa and Guanajuato, two other sources of large amounts of U.S. vegetable and fruit imports, respectively.

In Chile, a heavy contributor to the produce aisle's winter bounty in the United States, water troubles abound. As of fall 2019, the country's central zone, which includes the key export-ag regions Valparaiso and O'Higgins, was locked in an "uninterrupted sequence of dry years since 2010 with mean rainfall deficits of 20–40%," according to a paper from Universidad de Chile researchers. They deemed the event a "Mega Drought" with "few analogues in the last millennia." Examining the complex array of factors driving the drought—such as climate change and fluctuations in El Niño patterns—the authors concluded, "We anticipate only a partial recovery of central Chile precipitation in the decades to come." Avocado production in the water-stressed Valparaiso region, mostly destined for European and Asian markets, has dramatically expanded in recent years despite the drought, driving conflict between large plantation owners and small-scale farmers growing for the domestic market, who complain their wells have gone dry. In 2019, U.S. fresh fruit and vegetable imports from Chile fell sharply, owing to the drought.

Overall, the United States is already a large importer of produce. As of 2017, more than half of U.S.-consumed fruit and about a third of fresh vegetables were grown in other countries, and the proportions are growing. Given the stresses on our current suppliers, we cannot reasonably expect to be able, as Saudi Arabia did, to import our way out of a collapse in domestic production.

So if the dominant U.S. produce-growing state faces severe and increasing water scarcity, and our main import suppliers do, too, where should U.S. eaters look for sustenance in the coming decades? How should farmers adjust? And what policy solutions should lawmakers—who theoretically have eaters' and farmers' well-being in mind—pursue? The same questions apply to the Midwestern Corn Belt. The region has emerged as a factory for the inputs that go on to produce industrial meat—and degraded and compromised one of the globe's richest stores of topsoil in the process, just as it comes under increasing attack from climate-related storms.

If one theme unites the crises I've teased out in this book, it's simplification. Goaded on by farm interests, the federal and California governments poured billions of dollars into building out irrigation infrastructure in the Central Valley over the (anomalously wet) twentieth century, enabling a hyperefficient agriculture industry there, undercutting fresh-produce growers in other regions. In the Midwest, an interlocking set of corporate oligopolies presides over a veritable empire of meat. The result is a reassuringly simple geography of food—and one quite profitable for the corporations that dominate it: California largely provides our fruits and vegetables, the Midwest our meat.

But the ecologies that have supported these massively productive enterprises are unraveling.

It's time to mix them up and spread them out. In the Midwest, that means broadening the crop mix beyond corn and soybeans, and moving animals from feedlots and confinement facilities and back onto the land. For California, that will mean a scaling back of agriculture to fit the state's changing water resources. Both efforts would result in less gross output within the region; the challenge would be to increase food production in other areas.

Paring back our reliance on California doesn't mean the elimination of farming there—not at all. With its Mediterranean climate, endowment of rich soils, and once-glorious rivers fed by the snow-topped (though increasingly less so) Sierra Nevada, the state will always be an agricultural powerhouse. But why must just a few clusters of water-stressed counties in a single state provide 81 percent of U.S.-grown carrots, 95 percent of broccoli, 78 percent of cauliflower, 74 percent of raspberries, 91 percent of strawberries, 66 percent of lettuce, 63 percent of tomatoes, and on and on?

Since California is already committed to scaling back its agriculture to bring it in line with water realities—as will inevitably happen as the Sustainable Groundwater Management Act proceeds—U.S. lawmakers should consider putting public policy and resources behind a strategic ramp-up of produce farming in other regions outside California. A better idea for California's farms would be to turn away from focusing on vast monocrops for distant markets—salad greens trucked to the East Coast, almonds shipped to Beijing—and scale down to grow high-quality food for the arid Southwest's fast-growing cities, with long-distance exports soaking up the excess, instead of serving as the north star.

As for our meat habit, like it or not, we have to prepare for the coming drop in Corn Belt crop yields that will inevitably occur under pressure from hot temperatures, stressed crops, and declining soil quality. The factory-livestock farming system that requires such an enormous and inefficient allocation of corn and soybeans must be rethought from the ground up. The most viable solution is to ramp up pasture-based meat production, both within the Corn Belt and in regions across the country.

It's important to note that a widening distribution of farmed territory and a reorientation toward local needs have been taking place on a significant scale over the past twenty-five years. The number of U.S. farmers markets rose from less than 1,800 in 1994 to more than 8,000 by 2013. Community-supported agriculture projects—where consumers front money to buy a share in a farm's output, which is then distributed during the growing season—didn't appear in the United States until the early 1980s. By 2015, there were more than 7,300 farms relying at least partially on the CSA model, accounting for $226 million in sales, USDA figures show.

Overall, farm sales directly to consumers doubled between 1992 and 2007, reaching $3 billion by 2015, and holding steady since. As direct sales have flattened, locally grown food sold to wholesale markets and passed through to supermarkets and institutions like schools and hospitals has boomed, too, reaching $9 billion in 2017.

The nation's coastal metropolises, with their burgeoning local-food scenes, are major drivers of the trend. But the phenomenon is robustly national. Iowa itself boasts 220 farmers markets. The weekly downtown market in Des Moines features more than fifty vendors of fresh produce alone. Locally grown produce tends to be more expensive than the stuff trucked in from vast California operations, but it by no means appeals only to the well-off. Between 2004 and 2013, the number of markets that accept electronic payments from shoppers using federal food-aid programs shot up from 500 to more than 4,000, and sales through those channels have boomed. By 2014, in a 430 percent increase from 2006, 4,322 school districts had farm-to-school programs, bringing local produce into the National School Lunch Program, which serves mainly kids from low-income families. The USDA's map depicting the locations of farmers market by state looks like a population density map. In short, wherever people live, demand for food grown nearby has surged in the past twenty-five years.

The dramatic buildup in local food that has taken place in the twenty-first century is a priceless asset as we navigate the scaling back of California vegetable production and the unwinding of the Midwest's corn-soybean-meat complex. That's because the small and midsize farms that have arisen nationwide to supply booming new markets are a diverse lot. Some specialize in fresh produce; some focus on meat and dairy; others combine both.

And yet, standing amid the clamor, the scents, and the conversations of a typical vibrant farmers market, it's easy to forget the massive built environment for industrial food in the surrounding areas: fast-food outlets by the hundreds of thousands, supermarkets hawking aisle after aisle of processed food, school and hospital cafeterias—despite the above-mentioned farm-to-school push—still serving up not fresh-cooked fare but rather heat-and-serve fare, to the very folks who could use a good meal the most. Despite the farmers market revolution of the past quarter century, the larger food system still exists to transfer cheap meat and corn and soybean derivatives from the Midwest, supplemented with mass-produced vegetables shipped in from California and other countries, into our bodies.

Nationwide, farmers markets and other local-food institutions—for all their vibrancy, for all the efforts to broaden access to them—generally remain niche operations catering to a relatively small portion of the population. The boom in local food sales suggests a widespread yearning for something different; its leveling off in recent years suggests a blockage, something holding it back. To develop truly robust local and regional networks that can maintain a steady supply of fruit and vegetables, as California's water shocks mount and Iowa's soil nears depletion, we'll have to figure out why.

One reason: although the farmers market model can work for farms small enough to sell all or most of their produce directly to consumers, it makes only limited economic sense for diversified midsize family farms. As a group of agriculture scholars, led by Fred Kirschenmann, wrote in a seminal paper titled "Why Worry About Agriculture of the Middle?," midsize farms get squeezed in this arrangement. They are "too small to compete in highly consolidated commodity markets, and too large . . . to sell in direct markets." National distributors stocking huge grocery chains have the leverage to push down prices paid to farmers, and they can put midsize farmers in a particular region into competition with massive California or foreign operations.

And yet, Kirschenmann shows, it's precisely these midsize farms that have the scale to grow local and regional food chains to a point where they supply a large part of the American diet. For the past several decades, midsize farms have operated under severe pressure; they close at a rate of more than 10 percent every five years. Meanwhile, the number of large and mega-farms—the kind that dominate in the Corn Belt and California's main valleys—grows steadily. Finding economic models that allow midsize producers to thrive is crucial to the needed transformation of our food system.

David Swenson, an Iowa State University agriculture economist, has laid out one such model that could work in the heart of the Corn Belt, in ways that would benefit the region's rural and urban residents alike. Currently, the region's farmers—with their tens of millions of acres of prime farmland—essentially leave the fruit and vegetable market to distant peers, even during the warm months. But the Midwest's metropolitan areas represent a massive assembly of consumers. Combine Chicagoland (population 9.5 million), with greater Detroit (4.3 million), Minneapolis–St. Paul (3.3 million), St. Louis (2.8 million), and the others, and the region's metro zones house more than 35 million people—a level not much lower than California's itself (39 million). Today, despite the explosion in farmers markets (and even urban farms) within these sprawling metroplexes, California's agriculture hubs remain the dominant supplier of fruits and vegetables year-round.

The soil-rich Midwest's reliance on trucked-in food is a relatively recent habit. In a 2001 paper, Iowa State University researchers crunched historic data from U.S. Agricultural Census records and found that in 1920, Iowa was a bit of a hotbed of commercial fruit production; most farms, it turned out, devoted at least a small patch to it. That year, they reported, 84 percent of Iowa farms grew apples, 62 percent grew potatoes, more than half grew potatoes and cherries, and nearly a third grew plums and grapes. Other relatively common crops included strawberries, pears, and peaches. By 1997, not a single fruit or vegetable crop was grown on more than 1 percent of farms. By 2000, they concluded, Iowa farmers "grew a small fraction of the fresh produce bought during the summer months by Iowa produce wholesalers and distributors for sale within the state."

Similar patterns play out throughout the Corn Belt: fruit and vegetables trucked into a region that houses one of the globe's great stores of soil. Couldn't farms there increase production and pick up slack from water-challenged California? The answer is yes, with caveats.

It's important to understand that vegetable farming could ever take up only a relatively small portion of Corn Belt land. Crops like corn, soybeans, wheat, and oats are stable, in that they can be grown on a grand scale and stored cheaply and compactly until they're needed. The Midwest houses extensive infrastructure for storage (i.e., grain elevators), as well as for shipping them down the Mississippi for export, and for processing into everything from pork (in those hog CAFOs), to ethanol, to sweeteners and other ingredients. Horticulture crops such as kale and tomatoes are highly perishable, and extending their shelf life means costly processing and packaging. Storage requires large amounts of refrigerated space, including, potentially, freezers. A few billion bushels of unsold corn at the end of the season is a normal annoyance; a million pounds of unsold spinach is a potential economic disaster.

So expanding vegetable and fruit production isn't a silver-bullet solution to the Corn Belt's ecological crisis. But doing so at an appropriately modest scale is a smart idea, as Swenson showed in a 2010 paper.

He found that by moving just 270,000 acres of land (equivalent in size to a single typical Iowa county) from corn and soybeans to veggies, farmers in the relatively water-rich Midwest could supply everyone in Illinois, Indiana, Iowa, Michigan, Minnesota, and Wisconsin with half of their annual tomatoes, strawberries, apples, and onions, and a quarter of their kale, cucumbers, and lettuce. The economic advantage of such a modest transfer of land from commodity crops to food crops would be stunning, they found: it would generate $882.44 million in farm-level sales, which would be worth $3.31 billion when sold at retail. And it would create more than six thousand new, mostly rural jobs, delivering an additional $345.12 million in labor incomes, compared with the equivalent amount of land devoted to largely mechanized corn and soybean production.

In 2019, Swenson told me that his economic analysis had drawn few takers since its publication nearly a decade before. "It's one thing to show in a paper that a very different kind of farming can be profitable," he said. Convincing Corn Belt farmers to devote parts of their farms to perishable edibles is daunting. Per acre, a field of tomatoes, berries, or greens can generate much more income than an acre of soybeans. But it also requires entirely different skills and equipment, and a lot more labor. It didn't help, he added, that corn and soybean prices plunged in 2013, staying low through 2019. While low prices might be expected to spur farmers to try something, in practice the opposite is true: they can feel pressure to hunker down and to avoid new risks.

And even when crop prices slide, land prices stay stubbornly high—a reflection of federal crop insurance and subsidy programs for corn and soybeans that ensure Corn Belt farms stay afloat even during times of overproduction. In a business that operates on razor-thin margins, there's tremendous pressure to scale up by buying more land, hoping that selling a whole lot of corn at a tiny profit per bushel will add up to a decent profit on the whole. As a result, even during bad times, a large-scale corn and soybean grower looking to expand can typically outbid a young farmer hoping to grow, say, pastured meat and vegetables for Des Moines residents.

Swenson's grand vision is hardly the only fresh idea that has struggled to break through the inertia. Farmers like Ohio cover crop enthusiast David Brandt and Iowa rye proselytizer Tom Frantzen—and many more like them—prove the important concept that biodiversity can free the region from agrichemical domination, while still churning out high yields.

But while cover crops are expanding in popularity, they're used on a relatively tiny percentage of the region's land. Corn and soybeans still hold sway; rye and oats remain on the fringe, vestiges of the Corn Belt's past, still not established as a guide to its future. And the plunge in crop prices that began in 2013 has persisted since. Scraping by mainly with the aid of various government programs, farmers have little incentive to gamble on new practices, even when outliers among them prove they work. As for the cripplingly high prices for fertilizer and tractor fuel Brandt thought might inspire his neighbors to search for less chemical-intensive methods, that didn't pan out either. Fertilizer prices didn't fall as rapidly or steeply as crop prices, but they did fall—due to an ongoing domestic boom in the hydrofracturing (fracking) of petroleum and natural gas.

Meanwhile, the massive corporations that reign over the agriculture industry tap into multibillion-dollar markets for their goods. They command billions more in infrastructure designed to keep churning those goods out. They will defend these assets to the last dollar and invest some of their profits to curry favor in Washington, D.C., Des Moines, Sacramento, or any other political power center that could challenge them.

Over the years, talking to farmers and advocates in the Corn Belt about the stubborn grip of corn, soybeans, and confined hogs, I've heard over and over again about federal policy that encourages farmers to maintain the status quo. From direct payments to subsidized crop insurance, the system that has evolved since the 1980s keeps farmers just scraping by while delivering as much corn and soybeans as they can, typically way more crop than the market demands. So-called conservation programs, which pay farmers for practices like cover crops, are funded at pennies on the dollar, compared with these production incentives. They are like a weak set of brakes on a massive truck speeding toward a cliff. These policies—rolled over and ratified like clockwork every five years in the omnibus legislation called the Farm Bill—keep the profits humming for the agribusiness industry, which in turn uses its financial muscle to keep it that way.

As a whole, the agribusiness sector dishes out around $100 million in campaign contributions every two-year federal election cycle, reports the Center for Responsive Politics (CRP), which tracks corporate political spending. That's less than what's spent by the finance and health sectors, but more than defense, construction, and transportation. The U.S. House and Senate agriculture committees, which shape U.S. farm policy and prop up the corn-soybean duopoly through the Farm Bill, are essentially machines for sucking in agribusiness campaign donations. Senate ag committee members collectively draw around $9 million annually from agribusiness companies and their employers, while their House counterparts bring in around $8 million.

Beyond that, the agribusiness industry spent nearly $2.5 billion between 1998 and the first half of 2019 on lobbyists in the U.S. capital, according to CRP. Again, that's considerably less than the finance and health industries, but enough to place it sixth on the CRP's list of big-spending sectors, beating out defense contractors ($2.4 billion) and construction firms ($950 million).

When Monsanto and Bayer agreed to merge in 2016, the two companies had to convince antitrust regulators to approve the creation of a new seed-pesticide behemoth. It surely didn't hurt that the two companies had spent a combined $120 million on lobbying in the previous nine years. Post-merger, Bayer continues bombarding Washington with cash, dropping $5 million on lobbying in the first half of 2019.

Its rival DowDuPont followed a bolder path. While the merger between the two chemical giants was still pending in 2016, Dow contributed $1 million to newly elected president Donald Trump's inaugural committee. Around the same time, Trump named Dow CEO Andrew Liveris chair of the American Manufacturing Council, declaring the chemical exec would "find ways to bring industry back to America." In 2017, Dow achieved two major lobbying goals in Washington. That March, the Environmental Protection Agency reversed an Obama-era decision to ban a widely used insecticide called chlorpyrifos, a potent neurotoxin known to harm cognitive development in children at low exposure levels. Then, in June, the U.S. Department of Justice approve Dow's merger with DuPont.

The goodwill between the Trump administration and DowDuPont continued. In 2018, Trump named Scott Hutchins, Dow's longtime global director for crop protection (read: pesticides) R&D, to the powerful post of chief scientist at the U.S. Department of Agriculture, which oversees the agency's $2.9 billion research budget. Trump also put two former Dow men, including its onetime chief lobbyist, in charge of the USDA's foreign-trade agency. Then, in 2019, the combined DowDuPont spun out its agribusiness division as a stand-alone corporation called Corteva Agriscience, which is the world's second-biggest seed-agrichemical company after Bayer. Newly hatched Corteva immediately joined the influence game, spending $851,000 on lobbying in 2019 through July, mostly pushing for pesticide deregulation. Given the immense financial investment in keeping things the way they are, it's no wonder that the Corn Belt still looks the way it does: row after Technicolor-green row of corn and soybeans stretching out to the horizon in every direction, a seemingly boundless sameness gradually leaching the life from what was once some of the most productive land on the planet.

The Central Valley and the Corn Belt have supplied our shopping carts and kitchen tables at least for as long as anyone under seventy has been alive, with enough left over for massive exports to well-off nations worldwide. The setup has proved robust. When these two regions have suffered shocks, our food supply has hardly registered a blip—so far.

California's epic 2011–17 drought, compounded by unusually hot summers, essentially cut off Central Valley farmers from the snowmelt-fed irrigation projects they had relied on for decades. They responded by tapping deep into aquifers and prioritizing crops that have a high value in national and global markets, like almonds and the vegetables that fill U.S. supermarkets. Regional products like dairy, and nonfood ones like cotton, lost out. Farmers took a huge hit, incurring nearly $500 million for each drought year in additional water-pumping costs alone. Overall costs to farmers, including fallowed land, reached $1.5 billion per drought year. But because of tight competition from large-scale, export-minded farms in Mexico, they had little leverage to pass those costs on to consumers, so food prices barely budged.

As for the Corn Belt, if we want a hint of how lower yields driven by soil erosion and climate change will play out, we can look to a crisis generated by the U.S. government in the early aughts. The federal ethanol program can be seen as a kind of simulation: What would happen if, every year for the foreseeable future, we literally burned a third of the U.S. corn crop, a loss of about 10 percent of all corn grown worldwide? President George W. Bush maintained a zealous commitment to denying climate change over the course of his presidency, but he did make one concession to "clean energy": declaring his intent to end the U.S. "addiction to oil," he oversaw a dramatic expansion in government-mandated and subsidized ethanol production. In 2000, the year before he took power, less than a billion bushels of U.S. corn were converted into ethanol, about 5.5 percent of the corn crop. In 2008, his last year in power, ethanol sucked in 3.7 billion bushels, or 30 percent of the crop. By 2010, ethanol was claiming around 5 billion bushels annually. That level has held steady ever since, congealed into law by the Energy Act of 2005.

The move did little to cut U.S. energy consumption. But the price of corn jumped from less than $2 per bushel to peak (briefly) at over $7, despite a series of bumper harvests and an expansion of corn acres as farmers scrambled to take advantage of high prices. Markets for other globally traded farm commodities tend to follow suit when there's a change in a big one like corn. In the Corn Belt, farmers responded to the price surge by devoting more land to corn and less to soybeans, which drove up soybean prices. In the Great Plains, farmers switched from wheat to corn to take advantage of the windfall; that meant less, and thus more expensive, wheat. Changes inspired by the ethanol program cascaded through global food markets.

It wasn't just burning a third of the U.S. corn crop in our gas tanks that incited the price spike. Wall Street hot money, fleeing the real estate collapse, plowed into agricultural investments like corn futures, puffing up the boom and causing prices to gyrate at the whim of traders. Droughts in major growing regions—including Australia, a key wheat and rice producer—added pressure. Amid this hostile situation, the unprecedented drought and heat wave settled upon the Corn Belt in 2012, causing corn and soybean yields to plunge.

Crop prices spiked anew, exacerbating a food crisis that pushed millions into hunger around the globe. That's because in less-developed countries, grains such as corn are primarily consumed not by livestock (as in the United States) but by people—and one third of the globe's population, including countries in sub-Saharan Africa, Southeast Asia, and Latin America, rely on corn as a staple. In Mexico, the first spike in global corn prices sparked by the U.S. ethanol boom triggered a jump in tortilla prices and inspired nationwide "tortilla riots." The crisis also sparked food riots in Egypt, Mozambique, and Niger. By 2011, the global food crisis that began in 2006 had pushed 44 million people worldwide into poverty.

But in the United States, the effect was relatively muted. Here, corn and soybeans are industrial inputs: raw material for an assembly line geared to churn out cheap food. Consider a box of cornflakes, perhaps the most corn-centric product in the supermarket. In a 2008 report assessing the effect of ethanol on food prices, a U.S. Department of Agriculture researcher crunched the numbers to determine how much money companies spend on corn to make your cereal. The report found that an eighteen-ounce box of cornflakes contains about 12.9 ounces of milled corn. At $2.28 per bushel—the pre-crisis twenty-year-average price—the cereal maker paid all of 3.3 cents per box for corn. At the peak price of around $8 per bushel reached in 2012, the corn cost reached 11.6 cents. If companies added the entire increase onto the consumer, the hit would be less than a dime per box. An eighteen-ounce box of cornflakes contains eighteen servings—meaning that much-pricier corn translated to an increase of about a half-penny per bowl. For a family of four that consumes four bowls of cornflakes every day, that's about $9 extra for the year.

Obviously, even slightly higher food prices fall hard on the tens of millions of Americans who live below the poverty line. Between 2006 and 2012, rising food prices coincided with the worst recession since the Great Depression, each compounding the effect of the other for the poorest families. But it's fair to say that most Americans barely noticed the shock to corn and soybean prices that began with the ethanol boom starting in 2006, while millions of people in the developing world were harmed by it. As Iowa State's Rick Cruse noted, the U.S. price run-up played a role in triggering the turbulent Arab Spring.

Ultimately, Wall Street hot money fled commodity crops like corn even as Corn Belt farmers churned out a succession of bumper crops. By 2014, corn prices had returned to earth, and farmers were once again losing money while growing by far the world's biggest corn crop.

So two severe recent shocks—the Central Valley's epochal drought and the Corn Belt's turn to ethanol, which was compounded by a severe drought—passed without much impact on most Americans' plates. That should not give us a sense of false comfort; as any decent financial adviser will tell you, past performance is no guarantee of future results. The Corn Belt's ongoing cycle of massive overproduction and unchecked soil loss, compounded by climate change, leaves it more vulnerable to future droughts; one equal in severity to the 2012 dry spell will cause more damage as the region continues to shed topsoil. And spring seasons equal in severity to that of 2019 will cause ever-bigger floods and gully washouts, leading to more erosion and, again, more vulnerability.

Similarly, in California's next big drought, farmers will turn to aquifers still depleted from the last one. They will have to bore underground for ever-more expensive, salt-laced water, causing the land to sink further, and snarling up the very irrigation infrastructure that conveys water to their farms in good years. They'll also be increasingly encumbered by the Sustainable Groundwater Management Act. SGMA could go a long way toward preserving the valley as a key node of the U.S. food system, but at the cost of shrinking the region's growing capacity. A 2019 study by the Public Policy Institute of California found that to comply with SGMA, the San Joaquin Valley alone will likely have to fallow 780,000 of its 5.5 million acres of farmland, a loss of 15 percent of current capacity. But if over-pumping continues apace until 2040, the valley will be facing a massive reckoning anyway.

To feed us, the two regions' remarkably productive farmers are drawing down what are essentially fossil resources, at a rate much faster than they can be replenished, and in a way that makes them increasingly vulnerable to weather shocks. An ever-warming climate will put ever more pressure on California's vanishing groundwater and the Corn Belt's diminishing soil. The day looms when we as a public won't be spared the effects of these epochal drawdowns.

In the spring of 2006, on a book tour for his blockbuster seminal critique of the food system, *The Omnivore's Dilemma*, Michael Pollan visited the Union Square Greenmarket in Manhattan. Standing amid the throngs of New Yorkers lining up to buy gorgeous vegetables and pasture-raised meats grown on nearby farms, Pollan trained his gaze across Fourteenth Street, to a gleaming Whole Foods that had recently opened. It, too, was packed. For Pollan, the green market and the organic supermarket embodied two strands of an emerging new paradigm for food, one that would soon trounce the existing order. The experience felt "a little like a visit to Paris in the spring of 1968 must have felt, or perhaps closer to the mark, Peoples Park in Berkeley in the summer of 1969," Pollan wrote. "Union Square, which 75 years ago served as the red-hot center of the labor movement, is now, at least symbolically, ground zero of the food movement," he observed. "The market for alternative foods of all kinds—organic, local, pasture-based, humanely raised—represents the stirrings of a movement, or rather a novel hybrid: a market-as-movement." He argued that in the realm of food, citizens have unique power to make a difference through consumption choices:

> Whatever your politics, there are activities your tax money supports that I'm sure you find troublesome, if not deplorable. But you can't do anything about those activities—you can't withdraw your support—unless you're prepared to go to jail. Food is different. You can simply stop participating in a system that abuses animals or poisons the water or squanders jet fuel flying asparagus around the world. You can vote with your fork, in other words, and you can do it three times a day.

Despite impressive innovation among farmers and loads of conscientious fork-voting among consumers in the years since, things look considerably bleaker than they did on the blissful dawn just after *The Omnivore's Dilemma*'s publication. Farmers markets like the one at Union Square continue to boom; and Whole Foods is so established that the world's everything-store, Amazon, has seen fit to acquire it as a foothold for taking the grocery industry by storm. But the industrial food system that Pollan so eloquently decried remains dominant, continuing to provide the great bulk of our sustenance.

And even conscientious food consumerism has taken a corporate-friendly turn. As I write this in January 2020, a faux-meat product called the Impossible Burger has emerged as a sensation. Available at Burger King, Hardee's, and other chains nationwide, the Impossible offers consumers the flavor they expect from a fast-food burger along with the comforting knowledge that they're avoiding feedlot beef. But the alternative to the status quo it offers is limited. The Impossible Burger's main ingredient is soy protein concentrate—meaning that it relies on the same supply of Corn Belt crops that the meat industry does. The revolution we need likely will not come through the drive-through window at Burger King.

Meanwhile, the global concentration of atmospheric CO_2 has risen from 380 parts per million in 2006 to more than 410 parts per million. In 2018, the United Nations' Intergovernmental Panel on Climate Change issued a stark warning: Unless greenhouse emissions fall by about 45 percent by 2030, a world of climate chaos awaits—and that translates to the ever-fiercer droughts and storms in California and the Midwest. As the crisis mounts, a climate change denier holds the White House, and the EPA and the USDA are shot through with former oil and agribusiness industry execs and flacks. The possibility of pushing U.S. agriculture in radically new directions seems ever-more remote, even as the need to grows more dire.

We have reached the limits of "market-as-movement" to transform the food system. In an economy marked by severe inequality, wage stagnation, and persistently high levels of poverty, the market approach excludes the population that can't afford to pay the higher prices of organic, local food or devote time to cooking from scratch. It threatens to create an ecologically robust niche for the prosperous within a dominant food system that quietly lurches on, wrecking the environment and making people sick. Well-off consumers *should* vote with their forks three times a day, but the pace of positive change they create has been no match for Big Food's massive inertia and the rapid advance of climate change.

But something important has emerged over the same period: the stirrings of a new form of movement politics. In the wake of the 2007–08 financial meltdown, the worst since the Great Depression, new space has opened for questioning the political economy that props up a food system fueled by the destruction of finite soil and water resources; and that intensifies wealth inequalities that make cheap industrialized food an attractive option. In 2016, a young political activist/bartender from the Bronx, New York, shocked the political establishment, trouncing a powerful business-as-usual Democrat in a U.S. House primary and ultimately winning the seat. A year after taking office, Representative Alexandria Ocasio-Cortez teamed up with a wizened colleague, Senator Ed Markey of Massachusetts, to roll out a call for the Green New Deal.

What inspires me about the Green New Deal, which as I write this has galvanized the progressive wing of the Democratic Party and the burgeoning climate movement, is its theory of politics. The Green New Deal document introduced by Ocasio-Cortez and Markey doesn't aim to be a set of policies ready to sail through the current Congress. Rather, it sets ambitious policy goals—including to achieve "net-zero greenhouse gas emissions through a fair and just transition for all communities and workers" and "to create millions of good, high-wage jobs and ensure prosperity and economic security for all people of the United States"—that are almost impossible to imagine in the current political economy.

As for food and agriculture policy, too, the resolution calls for a "10-year national mobilization" to "eliminate pollution and greenhouse gas emissions from the agricultural sector" by "supporting family farming," "investing in sustainable farming and land use practices that increase soil health," and "building a more sustainable food system that ensures universal access to healthy food."

But it's bluntly agnostic on what policies should be put in place to achieve these goals, stipulating that they should emerge "through transparent and inclusive consultation, collaboration, and partnership with frontline and vulnerable communities, labor unions, worker cooperatives, civil society groups, academia, and businesses." Filling in the blanks of what specific legislation is to come out of the deal is left up to the public—including to social movements.

This is an extremely unusual strategy for U.S. politicians. Normal legislation involves a congressperson proposing "politically feasible" or "passable" policies and then trying to push them through the lobbyist-dominated congressional process. With the Green New Deal, Ocasio-Cortez and Markey acknowledge that the standard strategy has stalled when it comes to climate change. What is politically feasible when climate deniers control the presidency and the Senate? What's passable when the fossil fuel and agribusiness industries shower Washington, D.C., with nearly a quarter billion dollars every two-year election cycle?

Ocasio-Cortez and Markey's idea is to generate an upswell of grassroots support for policies that aren't feasible under President Donald Trump and Senate Majority Leader Mitch McConnell—but might soon be if the public mobilizes to demand it.

As I write this, details of what Ocasio-Cortez's push might look like for food and agriculture policy are only starting to emerge. The clearest plan put forward yet lies in the Green New Deal proposal laid out by Senator Bernie Sanders, released in August 2019. It includes an ambitious plan to remake the food system—and addresses the most dire problems discussed in this book.

In chapter 4, I laid out the gauntlet of oligopolies that dominate agriculture. Sanders's plan vows to "break up big agribusinesses that have a stranglehold on farmers and rural communities." I also mentioned how those companies lock farmers in perpetual cycles of overproduction, which leads to abuse of soil, chronic low crop prices, and steady profits for input suppliers and meatpackers. Sanders has countered with a policy mechanism ripped from the original New Deal: supply management, essentially a government program that helps farmers of big commodity crops like corn and soybeans coordinate planting decisions to avoid growing too much or too little. Along the same lines, he'd also reestablish a national grain reserve, a lapsed New Deal institution that collected excess crops in bountiful years to keep prices from plunging, and released them in bad years to avoid shortages (a notion that, as Sanders points out, makes lots of sense in an era of climate chaos).

That same month, Sanders's rival for the Democratic nomination, Senator Elizabeth Warren, laid out an equally ambitious agriculture platform, also calling for supply management and promising to "make agribusinesses pay the full costs of the environmental damage they wreak by closing the loopholes that CAFOs use to get away with polluting and beefing up enforcement of the Clean Air and Clean Water Acts against them."

Before Sanders and Warren, no major candidate since Jesse Jackson, who ran during the height of the brutal farm crisis in the 1980s, had proposed supply management. And for understandable reasons, the dominant seed, pesticide, fertilizer, grain-trading, and meat companies hate supply management because it puts the brakes on production and squeezes their profits. To reinforce the end of overproduction as a policy goal, Sanders's Green New Deal would subsidize farmers based on how much carbon they sequester in their land, not how much unneeded corn they produce—a move that would encourage the spread of soil-building practices like those David Brandt practices.

The platform doesn't mention California's water crisis directly, or the challenge it presents to fruit and vegetable availability nationwide. But it does present a remedy: Sanders promises to "connect consumers with local farms and healthy foods" through a variety of public investments, including cooperatively owned grocery stores and food-processing infrastructure, because "rampant consolidation in processing has led to a lack of facilities for small-scale, local producers."

It's impossible to know whether the Green New Deal strategy will work. No matter who wins the presidency, the transformational policies outlined in Sanders's and Warren's proposals are just words on the page without a mass mobilization—a genuine grassroots political movement to bolster the market-as-movement Pollan described in 2006. Absent that, we're likely heading into a hotter, less stable future with a food system that's as durable as an ice cube dropped on a sunny street. Vote with your fork, yes. But also, vote with your feet.

ACKNOWLEDGMENTS

Many thanks to Alice Brooke Wilson, my first reader and smartest critic, who has opened new ways of thinking and always pushed me to put power structures at the center of my analysis. Her support for this project, from germination to copyediting, has been immeasurable. This book could not have been written without the support of a loving family. In addition to Alice Brooke, I acknowledge my niece Whitney Walker Philpott, a steadying presence during tumultuous times; my siblings Sara, Paul, and Vassar, and brother-in-law Tyler Bell, with whom I've been through so much. My ace companions Hazel the Cat (2001–2019) and Lindy the Dog provided emotional support and amusement every day during the writing of this book. My beloved mother, Anne Sugrue (1941–2019), taught me everything about perseverance and commitment; my father, Tom Philpott Sr. (1942–1991), taught me to take the past seriously, question arbitrary authority, and stick up for the underdog. Almost everything I know about farming comes from the opportunities afforded me and the direct teachings of my North Carolina extended family, especially Bill Wilson, Bettie Morris, Hillary and Worth Kimmel, Carolyn Ashburn, Chuck Heron, and my many coworkers over the years at Maverick Farms. Special shout-out to the gardeners at the GreenThumb Hollenback Community Garden in Clinton Hill, Brooklyn, circa 2001–2003, for demonstrating that other worlds are indeed possible; and to my partners in podcasting and critical thought about the food system, Raj Patel and Rebecca McInroy. In Austin, I depended on help from friends Jose and Olga Garcia to keep things running smoothly around the house, allowing me to write.

To understand what goes on in the Central Valley, I required a team of Virgils: Matt Black, Rosanna Esparza, Janaki Jagannath, Dvera Saxton, Gustavo Aguirre, the Masumotos, and Joe Del Bosque; and the scientific work of B. Lynn Ingram, Frances Malamud-Roam, Jay Famiglietti, Donald Swain, and many others. Mark Arax's books and articles helped shape my understanding of an endlessly fascinating region. In Iowa, my guides were the Iowa Citizens for Community Improvement crew, including Jess Mazour, Hugh Espey, and Nick Schutt; Iowa State University stalwarts Richard Cruse and Matt Liebman; and farmers Denise O'Brien, Tom Frantzen, George and Patti Naylor, and Francis Thicke. Special thanks to 11th Hour Project and the director of its food and agriculture program, Sarah Bell; without their generous support and patience, I would not have had time to write and research this book. Special thanks as well to the team at *Mother Jones*, especially Kiera Butler, Maddie Oatman, Clara Jeffery, and Monika Bauerlein. At two crucial points in the process of writing this book, I stayed at the Mesa Refuge writers' residence in Point Reyes, California. I couldn't have started, much less finished, this project without that enormous privilege. I greatly appreciated the work of my agent, Kari Stuart, in finding a home for *Perilous Bounty.* And what a home she found. The crew at Bloomsbury USA have been a model of patience and care. Ben Hyman delivered a sensitive and nuanced edit. M. P. Klier provided masterful copyediting. Laura Phillips gracefully stewarded the book through production on a short timeline.

Any errors and misconceptions in *Perilous Bounty* belong to me; anything valuable and interesting in it emerges from a life lived among smart, challenging people, and reading smart, challenging things.

NOTES

INTRODUCTION

1 **"hallucinogenic in their intensity"** Jim Leff, "Pretzel/Potato Chip Tasting, and Jim Meets Chickens," Chowhound, September 15, 2006, https://www.chowhound.com/food-news/132/pretzelpotato-chip-tasting-and-jim-meets-chickens.

3 **Sexton went on to cite** Steve Sexton, "The Inefficiency of Local Food," Freakonomics, November 14, 2011, accessed November 27, 2019, http://freakonomics.com/2011/11/14/the-inefficiency-of-local-food.

4 **In 2015, at the drought's height** USDA Economic Research Service, "Data Products, Farm Income and Wealth Statistics, Annual Cash Receipts by Commodity," accessed November 29, 2019, https://data.ers.usda.gov/reports.aspx?ID=17832.

5 **The 2011–17 drought** "A Drier California Than Ever? Pretty Much," *Los Angeles Times*, February 3, 2014, accessed November 10, 2019, https://www.latimes.com/nation/la-oe-ingram-california-drought-20140203-story.html.

CHAPTER 1: HIGH AND DRY

14 **Statewide, the same forces . . . 22,500 acres in 2006** Brenna Aegerter et al., *Asparagus Production in California*, UC Vegetable Research and Information Center, https://anrcatalog.ucanr.edu/pdf/7234.pdf.

14 **8,000 in 2018** Ching Lee, "Asparagus Farmers Work to Remain Competitive," March 21, 2018, *AgAlert*, http://www.agalert.com/story/?id=11696.

14 **One reason almond acreage doubled** California Department of Food and Agriculture, *2017 California Almond Acreage Report*, https://www.nass.usda

.gov/Statistics_by_State/California/Publications/Specialty_and_Other _Releases/Almond/Acreage/201804almac.pdf.

14 **Almond trees originated** Alan Davidson, *The Oxford Companion to Food* (London: Oxford University Press, 1999), 12. See also Bijan Jobrani, "The Fight to Save Iran's Ancient Almonds," *Subversify*, March 20, 2013, http://subversify .com/2013/03/07/the-fight-to-save-irans-ancient-almonds.

14 **It takes about a gallon of water** Alex Park and Julia Lurie, "It Takes How Much Water to Grow an Almond?!," *Mother Jones*, February 24, 2014, https://www.motherjones.com/environment/2014/02/wheres-californias -water-going. See also Mesfin M. Mekonnen and Arjen Y. Hoekstra, "The Green, Blue and Grey Water Footprint of Crops and Derived Crop Products," *Hydrology and Earth System Sciences Discussions* 8, no. 1 (2011): 1577–1600.

14 **California's one million acres of groves** California Department of Food and Agriculture, *2018 California Almond Acreage Report*, http://www.almonds .com/sites/default/files/content/attachments/2018%20Acreage%20 Report_USDA.pdf.

14 **consume four times more water** Julia Lurie, "California's Almonds Suck as Much Water Annually as Los Angeles Uses in Three Years," *Mother Jones*, January 2, 2015, https://www.motherjones.com/environment/2015/01/almon ds-nuts-crazy-stats-charts.

14 **what's essentially a desert** Christina Nunez, "Deserts, Explained," *National Geographic*, https://www.nationalgeographic.com/environment/habitats /deserts.

15 **average amount of snow** John C. Fyfe, Chris Derksen, Lawrence Mudryk, Gregory M. Flato, Benjamin D. Santer, Neil C. Swart, Noah P. Molotch, et al., "Large Near-Term Projected Snowpack Loss over the Western United States," *Nature Communications* 8, no. 1 (2017): 1–7. See also Philip W. Mote, Sihan Li, Dennis P. Lettenmaier, Mu Xiao, and Ruth Engel, "Dramatic Declines in Snowpack in the Western US," *NPJ Climate and Atmospheric Science* 1, no. 1 (2018): 1–6.

15 **projects a further loss** Fyfe et al., "Large Near-Term Projected Snowpack Loss over the Western United States."

16 **every 1-degree-Celsius rise** Laurie S. Huning and Amir AghaKouchak, "Mountain Snowpack Response to Different Levels of Warming," *Proceedings of the National Academy of Sciences* 115, no. 43 (2018): 10932–37.

16 **0.5-degree-Celsius jump** Intergovernmental Panel on Climate Change, "Summary for Policymakers of IPCC Special Report on Global Warming of 1.5° C Approved by Governments," 2018.

17 **California farms churn out** California Department of Food and Agriculture, "California Agricultural Production Statistics: 2018 Crop Year—Top 10 Commodities for California Agriculture," https://www.cdfa.ca.gov/statistics.

18 **"flowed uninterrupted into valleys"** Norris Hundley, *The Great Thirst: Californians and Water—a History* (Berkeley: University of California Press, 2001), 5.

19 **4 million acres of the Central Valley** Philip Garone, *The Fall and Rise of the Wetlands of California's Great Central Valley* (Berkeley: University of California Press, 2011), 2.

19 **salmon was so abundant** Ronald Yoshiyama, Frank Fisher, and Peyer Moyle, "Historical Abundance and Decline of Chinook Salmon in the Central Valley Region of California," *North American Journal of Fisheries Management* 18 (1998): 487–521.

20 **diet was plentiful and varied** William Preston, *Vanishing Landscape: Land and Life in the Tulare Lake Basin* (Manchester, NH: Olympic Marketing, 1981); population estimate: 39; diet: 35–36; landscape management: 46.

20 **Today's almond, pistachio, and grape plantations** Discussion of the Central Valley's precolonial ecology, as well as several quotations in this section, come from Hundley, *The Great Thirst*, 5–11.

21 **indigenous population had plunged** Preston, *Vanishing Landscape*, 57–88.

21 **essentially obliterated the native grass** Preston, *Vanishing Landscape*, 59.

22 **settlers ultimately moved** Preston, *Vanishing Landscape*, 82–83.

22 **reservation has a population** "Tule River Indian Reservation: Planning Program for the Master Plan" June 2015. https://web.archive.org/web/20171007013908/http://www.tulerivertribe-nsn.gov/wp-content/uploads/2016/09/Tule-Master-Plan-Administrative-Draft.pdf.

23 **about 380,000 acres of wetland** Garone, *Fall and Rise of the Wetlands*, 2, 4 (chart).

25 **Subsidence is the enemy** J. F. Poland, B. E. Lofgren, R. L. Ireland, and R. G. Pugh, *Land Subsidence in the San Joaquin Valley, California, as of 1972* (Washington, D.C.: Government Printing Office, 1975), https://pubs.usgs.gov/pp/0437h/report.pdf.

25 **California had created** Richard Walker, "It's Still Chinatown, Jake," *Brooklyn Rail*, November 5, 2015, https://brooklynrail.org/2015/11/field-notes/its-still-chinatown-jake.

26 **The water table plunged** B. Lynn Ingram and Frances Malamud-Roam, *The West Without Water: What Past Floods, Droughts, and Other Climate Clues Tell Us About Tomorrow* (Berkeley: University of California Press, 2013), 44–45; and California Department of Water Resources, "California's Most Significant Droughts: Comparing Historical and Recent Conditions," February 2015, pp. 23 and 39–47.

26 **dug into the San Joaquin Valley aquifers** Claudia Faunt and Michelle Sneed, "Water Availability and Subsidence in California's Central Valley," *San Francisco Estuary and Watershed Science* 13, no. 3, article 4, and chart on p. 6, https://escholarship.org/uc/item/6qn2711x#page-6. The figure takes into account the relatively modest groundwater recharge that occurs during wet years.

26 **entire city of Los Angeles** Los Angeles Department of Water and Power, *Briefing Book, 2017–2018*, p. 24, https://s3-us-west-2.amazonaws.com/ladwp-jtti/wp-content/uploads/sites/3/2017/09/08143247/Briefing-Book-Rolling-PDF.pdf. It reports the city uses on average 542,700 acre-feet per year, which equals 0.67 cubic kilometers.

27 **reduced the canal's maximum flow** NASA Jet Propulsion Laboratory, "NASA Data Show California's San Joaquin Valley Still Sinking," February 28, 2017, accessed November 6, 2018, https://www.jpl.nasa.gov/news/news.php?feature=6761. See also "Subsidence Shrinks Friant-Kern Canal Capacity by 60 Percent," *Western Farm Press*, January 9, 2018, accessed March 7, 2019, https://www.westernfarmpress.com/irrigation-systems/subsidence-shrinks-friant-kern-canal-capacity-60-percent.

28 **high-pressure concentration recedes** "Climate of California," Western Regional Climate Center, accessed December 29, 2018, https://wrcc.dri.edu/narratives/CALIFORNIA.htm; and Daniel L. Swain, Baird Langenbrunner, J. David Neelin, and Alex Hall, "Increasing Precipitation Volatility in Twenty-First-Century California," *Nature Climate Change*, April 23, 2018, accessed December 29, 2018, https://www.nature.com/articles/s41558-018-0140-y.

28 **bountiful water year** Michael Dettinger, Michael D., Fred Martin Ralph, Tapash Das, Paul J. Neiman, and Daniel R. Cayan, "Atmospheric Rivers,

Floods and the Water Resources of California." *Water* 3, no. 2 (2011): 445–78, March 24, 2011, https://www.mdpi.com/2073-4441/3/2/445.

29 **"ridiculously resilient ridge"** Donald Swain, "New Insights into the Ridiculously Resilient Ridge and North American Winter Dipole," *California Weather Blog*, December 4, 2017, accessed August 5, 2019, https://weatherwest.com/archives/5982.

29 **clearly increasing trend** Daniel L. Swain, Deepti Singh, Daniel E. Horton, Justin S. Mankin, Tristan C. Ballard, and Noah S. Diffenbaugh, "Remote Linkages to Anomalous Winter Atmospheric Ridging over the Northeastern Pacific," *Journal of Geophysical Research: Atmospheres* 122, no. 22 (2017), 12,194–209; and Daniel L. Swain, Daniel E. Horton, Deepti Singh, and Noah S. Diffenbaugh, "Trends in Atmospheric Patterns Conducive to Seasonal Precipitation and Temperature Extremes in California," *Science Advances* 2, no. 4 (2016): e1501344.

30 **"increased soil drying"** P. A. Ullrich, Z. Xu, A. M. Rhoades, M. D. Dettinger, J. F. Mount, A. D. Jones, and P. Vahmani, "California's Drought of the Future: A Midcentury Recreation of the Exceptional Conditions of 2012–2017," *Earth's Future* 6, no. 11 (2018): 1568–87.

30 **average annual snowpack** Alan M. Rhoades, Andrew D. Jones, and Paul A. Ullrich, "The Changing Character of the California Sierra Nevada as a Natural Reservoir," *Geophysical Research Letters* 45, no. 23 (2018): 13,008–19.

30 **result of projected warming** "Sierra Nevada Region Report—California's Fourth Climate Change Assessment," August 27, 2018, accessed May 24, 2019, http://climateassessment.ca.gov.

31 **"megadrought periods of the Medieval era"** Benjamin I. Cook, Toby R. Ault, and Jason E. Smerdon, "Unprecedented 21st Century Drought Risk in the American Southwest and Central Plains," *Science Advances* 1, no. 1 (2015): e1400082.

32 **"there is no drought"** "Trump Tells California 'There Is No Drought,'" CNBC, May 26, 2016, accessed March 4, 2019, https://www.cnbc.com/2016/05/28/trump-tells-california-there-is-no-drought.html.

32 **could well spell doom** Doug Obegi, "Trump's Bay-Delta Biops Are a Plan for Extinction," NRDC, November 11, 2019, accessed November 20, 2019, https://www.nrdc.org/experts/doug-obegi/trumps-bay-delta-biops-are-plan-extinction.

33 **arid western side of the valley** Paul D. Welle and Meagan S. Mauter, "High-Resolution Model for Estimating the Economic and Policy Implications of Agricultural Soil Salinization in California," *Environmental Research Letters*, September 2017, accessed November 8, 2018.

CHAPTER 2: THE FLOOD NEXT TIME

34 **Brewer disembarked** John Brewer, *Up and Down California in 1860–1864: The Journal of William H. Brewer* (Berkeley: University of California Press, 2003), 221–38.

36 **"*six feet*"** Brewer, *Up and Down California*, 241 [emphasis in original].

36 **"An old acquaintance, a *buccaro*"** Brewer, *Up and Down California*, 244.

36 **"Most of the city is still under water"** Brewer, *Up and Down California*, 249.

37 **"ecology of fear"** Mike Davis, *Ecology of Fear: Los Angeles and the Imagination of Disaster* (New York: Vintage Books, 1998), 5, 7.

38 **Cassandra of flood preparedness** B. Lynn Ingram and Frances Malamud-Roam, *The West Without Water: What Past Floods, Droughts, and Other Climatic Clues Tell Us About Tomorrow* (Berkeley: University of California Press, 2013), 30–31.

38 **As floodwater gathered** Jim Goodridge, "Data on California's Extreme Rainfall from 1862–1995," June 29, 1996, accessed December 7, 2018, https://cepsym.org/Sympro1996/Goodridge.pdf.

38 **remnant indigenous population** B. Lynn Ingram, "California Megaflood: Lessons from a Forgotten Catastrophe" *Scientific American*, January 1, 2013, accessed December 7, 2018, https://www.scientificamerican.com/article/atmospheric-rivers-california-megaflood-lessons-from-forgotten-catastrophe.

39 **"disaster capitalism"** Naomi Klein, *Shock Doctrine: The Rise of Disaster Capitalism* (New York: Picador, 2008, paperback edition).

39 **1862 flood** James Lawrence Jelinek, "'Property of Every Kind': Ranching and Farming During the Gold-Rush Era," *California History* 77, no. 4 (1998): 233–49.

40 **supine Sacramento** Brewer, *Up and Down California*, 250.

40 **population of 6.5 million** California Department of Finance, "New Demographic Report Shows California Population Nearing 40 Million Mark with Growth of 309,000 in 2017," May 1, 2018, accessed December 10, 2018,

http://www.dof.ca.gov/Forecasting/Demographics/Estimates/e-1/documents /E-1_2018PressRelease.pdf.

40 **high-speed train** Dale Kasler, Ryan Lillis, and Tim Sheehan, "California's Bullet Train Is Pumping Billions into the Valley Economy. So Why Is It So Unpopular?," *Sacramento Bee*, December 23, 2018, accessed December 29, 2018, https://www.sacbee.com/latest-news/article223441880.html.

40 **Great Flood of 1862** Michael Dettinger and B. Lynn Ingram, "Megastorms Could Drown Massive Portions of California," *Scientific American*, January 1, 2013.

42 **twenty-five times the flow** F. M. Ralph, S. F. Iacobellis, P. J. Neiman, J. M. Cordeira, J. R. Spackman, D. E. Waliser, G. A. Wick, A. B. White, and C. Fairall, "Dropsonde Observations of Total Integrated Water Vapor Transport Within North Pacific Atmospheric Rivers," *Journal of Hydrometeorology* 18, no. 9 (2017): 2577–96.

42 **pre-Columbian cataclysms** Dettinger and Ingram, "Megastorms Could Drown Massive Portions of California."

42 **lifeblood** Michael Dettinger, "Atmospheric Rivers as Drought Busters on the U.S. West Coast," *Journal of Hydrometeorolgy* 14 (2013): 1721–32, accessed January 2, 2019, https://doi.org/10.1175/JHM-D-13-02.1.

43 **atmospheric-river storms** Yong Zhu and Reginald E. Newell, "A Proposed Algorithm for Moisture Fluxes from Atmospheric Rivers," *Monthly Weather Review* 126 (1998): 725–35, accessed January 2, 2019, https://doi.org/10.1175 /1520-0493(1998)126<0725:APAFMF>2.0.CO;2.

43 **thirteen biggest storms** Thomas Corringham, F. Martin Ralph, Alexander Gershunov, Daniel R. Cayan, and Cary A. Talbot, "Atmospheric Rivers Drive Flood Damages in the Western United States," *Science Advances* 5, no 12 (2019), DOI: 10.1126/sciadv.aax4631.

43 **fickle transfers** Michael Dettinger, "Historical and Future Relations Between Large Storms and Droughts in California," *San Francisco Estuary and Watershed Science* 14, no. 2 (2016), https://escholarship.org/uc/item/8c5574hf.

43 **climate models** Daniel Swain, Baird Langenbrunner, J. David Neelin, and Alex Hall, "Increasing Precipitation Volatility in Twenty-First-Century California," *Nature Climate Change* 8, no. 5 (2018): 427.

45 **ARkStorm (for "atmospheric river 1,000 storm") Scenario** Keith Porter, Anne Wein, Charles N. Alpers, Allan Baez, Patrick L. Barnard, James Carter,

Alessandra Corsi, et al., *Overview of the ARkStorm Scenario*, Open-File Report No. 2010-1312, U.S. Geological Survey, 2011, https://pubs.usgs.gov/of/2010/1312.

45 **invest resources** "Dr. Lucile Jones Full Biography," accessed May 25, 2019, http://drlucyjones.com/wp-content/uploads/2016/02/DrLucyJones_Biography_2016.pdf.

46 **Hurricane Katrina** National Oceanic and Atmospheric Administration, "Billion-Dollar Weather and Climate Disasters: Table of Events, 1980–2019," accessed May 1, 2019, https://www.ncdc.noaa.gov/billions/events/US/1980-2019.

47 **Tulare Lake bottom** Geoffrey Plumlee, Charles Alpers, Suzetter Morman, and Carma San Juan, "Anticipating Environmental and Environmental-Health Implications of Extreme Storms: ARkStorm Scenario," *Natural Hazards Review* 17, no. 4 (2016), accessed May 27, 2019, https://ascelibrary.org/doi/abs/10.1061/%28ASCE%29NH.1527-6996.0000188.

47 **much more populous** Population in 1860: Joseph Camp Griffith Kennedy, *Population of the United States in 1860* (Government Printing Office, 1864). Current population: Ryan Lillis, "On the Rise? The Central Valley Is Beating the Bay Area and L.A. in Key Measures," January 11, 2019, accessed May 26, 2019, https://faithinthevalley.org/on-rise-central-valley-beating-bay-area-l-a-key-measures.

47 **5 million head of beef cattle** National Agricultural Statistics Service, "California Cattle County Estimates," USDA, May 15, 2017, accessed May 26, 2019, https://www.nass.usda.gov/Statistics_by_State/California/Publications/County_Estimates/2016/201705LvstkCActy.pdf.

47 **dairy industry** California Department of Food and Agriculture, "2017 Dairy Statistics Annual," accessed May 26, 2019, https://www.cdfa.ca.gov/dairy/pdf/Annual/2017/2017_Statistics_Annual.pdf.

48 **post-flood Sacramento** John Newbold, "The Great California Flood of 1861–1862," *San Joaquin Historian*, Winter 1991.

48 **500,000 dairy cows** California Department of Food and Agriculture, "2017 Dairy Statistics Annual."

48 **massive creatures** "History of the Holstein Breed," Holstein Association, accessed May 27, 2019, http://www.holsteinusa.com/holstein_breed/breedhistory.html.

48 **industrial chemicals** Plumlee et al., "Anticipating Environmental and Environmental-Health Implications of Extreme Storms."

48 **Kern's farmers** California Department of Pesticide Regulation, "Total Pounds of Pesticide Active Ingredients Reported in Each County and Their Rank During 2016 and 2017," https://www.cdpr.ca.gov/docs/pur/pur17rep/tables/table3.htm.

49 **200 million pounds of pesticides** "Pesticide Use Near Record High in California," Pesticide Action Network," 2018, accessed May 27, 2019, http://www.panna.org/press-release/pesticide-use-near-record-high-california.

49 **oil refineries** "Oil in Kern: Oil, Gas Play Key Role for Kern County Public Finances," Duke Energy Initiative, accessed May 27, 2019, https://energy.duke.edu/content/oil-gas-play-key-role-kern-county-public-finances. Refineries: "California's Oil Refineries," California Energy Commission, accessed May 27, 2019, https://www.energy.ca.gov/almanac/petroleum_data/refineries.html. Cows: USDA National Agricultural Statistics Service, "California Cattle County Estimates," May 15, 2017, accessed May 26, 2019, https://www.nass.usda.gov/Statistics_by_State/California/Publications/County_Estimates/2016/201705LvstkCActy.pdf.

49 **twice as much crude** "How Big Is California's Oil and Gas Industry?," OilPrice.com, June 11, 2017, accessed May 27, 2019, https://oilprice.com/Energy/Crude-Oil/How-Big-Is-Californias-Oil-And-Gas-Industry.html.

50 **California National Guard** Samantha Schmidt, Derek Hawkins, and Kristine Phillips, "188,000 Evacuated as California's Massive Oroville Dam Threatens Catastrophic Floods," *Washington Post*, February 13, 2017, https://www.washingtonpost.com/news/morning-mix/wp/2017/02/13/not-a-drill-thousands-evacuated-in-calif-as-oroville-dam-threatens-to-flood.

50 **slow-moving traffic** Tony Bizjak, "'Mass Chaos' of Oroville Evacuation Prompts Worry over Exit Strategy," *Sacramento Bee*, February 18, 2017, https://www.sacbee.com/news/local/transportation/back-seat-driver/article133485154.html.

51 **future cataclysm** Federal Energy Regulatory Commission, "Submittal of Lake Oroville Probable Maximum Flood Update, Spillway Recovery Project," October 25, 2018, https://www.circleofblue.org/wp-content/uploads/2018/11/FERC_20181025-Oroville.pdf.

51 **dam's safety** California Department of Water Resources, "Lake Oroville Community Update," October 21, 2019, https://water.ca.gov/News/Blog/2019/Oct-19/Oroville-Community-Update-October-18.

CHAPTER 3: PUMPING AIR

52 **household income** Alpaugh Community Plan 2017, Tulare County General Plan, accessed November 10, 2018, http://generalplan.co.tulare.ca.us/documents/GP/001Adopted%20Tulare%20County%20General%20Plan%20Materials/120Part%20III%20Community%20Plans%201%20of%207/001Alpaugh/GPA%2017-04%20ALPAUGH%20COMMUNITY%20PLAN.pdf.

52 **poverty rate** Data USA, Alpaugh, CA, accessed November 10, 2018, https://datausa.io/profile/geo/alpaugh-ca/.

54 **arsenic-filtration system** Bond Accountability, Reference Number: 3940P01D1702042, Alpaugh CSD Arsenic Treatment Construction Project, http://bondaccountability.resources.ca.gov/Project.aspx?ProjectPK=22101&PropositionPK=48.

55 **2,500 acres** "2015 Sample Costs to Establish and Produce Pistachios," University of California Cooperative Extension, accessed November 10, 2018, https://coststudyfiles.ucdavis.edu/uploads/cs_public/50/a8/50a8805e-03a8-4092-82ca-ce18cfb92b2b/2015pistachios_san_joaquin_valley-south_oct29.pdf.

57 **Wonderful Company** Wonderful Pistachios, "Press and Trade," accessed November 13, 2018, https://www.getcrackin.com/press.html.

57 **50,000 acres of nuts** "American and Western Fruit Grower's 2014 Top Nut Growers," Growing Produce, accessed November 13, 2018, https://www.growingproduce.com/nuts/2014-top-nut-growers.

57 **$1.5 billion-per-year pistachio crop** "Pistachio Sales Approaching Almonds in Snacking, Says Wonderful Exec," *Fresh Fruit Portal*, July 28, 2017, accessed November 13, 2018, https://www.freshfruitportal.com/news/2017/07/28/pistachio-sales-approaching-almonds-snacking-says-wonderful-exec.

57 **mandarin oranges** "Paramount Citrus Replaces Cuties Brand with Wonderful Halos," *The Packer* June 10, 2013, accessed November 14, 2018, https://www.thepacker.com/article/paramount-citrus-replaces-cuties-brand-wonderful-halos.

58 **80 percent of the world's almonds** Ashley Bloemhof, "California Almonds: USDA Estimates Record-Breaking 2018 Crop," *AgFax*, July 5, 2018, accessed

November 13, 2018, https://agfax.com/2018/07/05/california-almonds-usda-estimates-record-breaking-2018-crop.

58 **bulk of them in the San Joaquin Valley** Almond Industry Maps, Almond Board of California, accessed November 14, 2018, http://www.almonds.com/processors/resources/crop-forecast/almond-industry-maps.

58 **counterparts in Iran** USDA Foreign Agricultural Service, Pistachio Summary, accessed November 13, 2018, https://www.documentcloud.org/documents/5096481-USDAFAS-Pistachio-Summary.html.

58 **2.3 pounds per year** USDA Economic Research Service, Fruit Tree and Nut Yearbook, https://www.ers.usda.gov/data-products/fruit-and-tree-nut-data/fruit-and-tree-nut-yearbook-tables/#Tree%20Nuts. See also "Almonds—Supply and Utilization," accessed November 13, 2018, https://www.documentcloud.org/documents/5096605-Almonds-FruitYearbookTreeNuts-FTables.html.

58 **Pistachios have enjoyed** USDA Economic Research Service, Fruit Tree and Nut Yearbook; and "Pistachios—Supply and Utilization."

59 **70 percent of California's almond crop** *Almond Almanac 2017*, Almond Board of California, http://www.almonds.com/sites/default/files/2017%20Almanac%20Final%20-%20updated%207.5.18.pdf.

59 **nearly half of its pistachios** USDA Foreign Agricultural Service, "Pistachio Summary," accessed November 13, 2018, https://www.documentcloud.org/documents/5096481-USDAFAS-Pistachio-Summary.html.

59 **Indian market** "California Almond Facts: India," Almond Board of California, https://www.almonds.com/sites/default/files/2017gtra0032_india.pdf.

59 **Chinese consumers** "Taste the Sunshine—New Advertising in China Features Benefits of California Almonds," Almond Board of California, accessed November 13, 2018, http://newsroom.almonds.com/content/taste-sunshine-%E2%80%94-new-advertising-china-features-benefits-of-california-almonds.

59 **French women** "New Advertising Targets Women in France," Almond Board of California, accessed November 13, 2018, https://newsroom.almonds.com/content/new-advertising-targets-women-france.

60 **Wonderful exec got the crowd** Dennis Pollock, "Paramount Farms Touts Record Pistachio Return, Future," *Western Farm Press*, March 16, 2015, accessed November 14, 2018, https://www.westernfarmpress.com/tree-nuts/paramount-farms-touts-record-pistachio-return-future.

61 **deep pockets** *2016 Sample Costs to Establish an Orchard and Produce: Almonds*, UC Cooperative Extension, accessed November 27, 2018, https://coststudyfiles.ucdavis.edu/uploads/cs_public/87/3c/873c1216-f21e-4e3e-8961-8ece2d647329/2016_almondsjv_south_final_10142016.pdf; and *2015 Sample Costs to Establish and Produce Pistachios*, UC Cooperative Extension, accessed November 10, 2018, https://coststudyfiles.ucdavis.edu/uploads/cs_public/50/a8/50a8805e-03a8-4092-82ca-ce18cfb92b2b/2015pistachios_san_joaquin_valley-south_oct29.pdf.

62 **land's value keeps rising** "A Company Overview," Hancock Agricultural Investment Group, 2018, https://www.documentcloud.org/documents/5237453-HAIG-Q318-Final-1.html.

62 **NCREIF Farmland Index allows investors** "How does direct investment in global farmland provide diversification, inflation protection, and return potential?," Nuveen, 2019, https://www.nuveen.com/en-us/thinking/alternatives/investing-in-farmland.

62 **9.6 percent** "Low-Hanging Fruit: Why You Should Plant U.S. Agriculture in Your Institutional Portfolio," PGIM Real Estate Finance, accessed November 27, 2018, http://www.pgimref.com/real-estate-finance/pdf/ag/Charticle_PAI_Booklet_v2_.pdf.

62 **ban ownership of farmland** "The Investable Farmland Universe: 172 Million Hectares, Valued at $1.5 Trillion," Hancock Farmland Investor, August 2018, accessed November 29, 2018, https://www.documentcloud.org/documents/5331725-Hancock-Farmland-Investor-the-Investable.html.

63 **nut orchards and wine grapes** "Farmland Property Index," NCREIF, accessed November 26, 2018, https://www.ncreif.org/data-products/farmland.

63 **planting nut groves . . ."Yankee Stadium"** Heather Davis, "Almonds: Harvesting Value Beyond the Farm," TIAA-CREF, August 7, 2013, https://www.tiaa.org/public/pdf/C11815_AlmondHarvest.pdf.

63 **TIAA not only snaps up** Tim Hopper, Biff Ourso, and Bruce Sherrick, "Farmland Investing in California: Withstanding a Historic Drought," TIAA-CREF, accessed November 27, 2018, https://www.tiaa.org/public/pdf/C15680-WithstandingHistoricDrought.pdf.

63 **40 percent stake in Treehouse** Board of Governors of the Federal Reserve System, Annual Report of Holding Companies, TIAA-CREF, 2016, https://www.documentcloud.org/documents/5301983-Tiaa-Board-of-Overseers-3792687-2016-1.html.

64 **HAIG owns at least twenty-four thousand acres** Hancock Agricultural Investment Group Company Overview, 2018, https://www.documentcloud.org/documents/5237453-HAIG-Q318-Final-1.html.

64 **Prudential Financial** "Agriculture Equity Investments—PGIM Real Estate Finance," November 27, 2018, http://www.pgimref.com/real-estate-finance/businesses-agricultural-equity-investments.shtml.

64 **financialization of the U.S. farm** *Low-Hanging Fruit: Why You Should Plant US Agriculture in Your Institutional Portfolio*, PGIM Real Estate Finance, April 2019, http://www.pgimref.com/real-estate-finance/pdf/ag/Charticle_PAI_Booklet_v2_.pdf.

64 **Gladstone Land Corporation** "Farmland Portfolio—Gladstone—Land & Farm," accessed November 28, 2018, http://gladstonefarms.com/farmland-portfolio.

65 **two San Joaquin Valley orchards** "Gladstone Land Acquires Two Orchards in California for $6.9 Million," Gladstone Land news release, September 17, 2018, accessed November 28, 2018, http://ir.gladstoneland.com/news-releases/news-release-details/gladstone-land-acquires-two-orchards-california-69-million.

65 **Wall Street titans** "Gladstone Land Corporation: Major Holders" Yahoo! Finance, accessed November 28, 2018, https://finance.yahoo.com/quote/LAND/holders?p=LAND.

65 **grapes, oranges, and walnuts** "Investable Farmland Universe," Hancock Farmland Investor.

66 **85 percent of U.S.-grown carrots** "Carrots," Agricultural Marketing Research Center, https://www.agmrc.org/commodities-products/vegetables/carrots [as of February 2020].

68 **$15,000 per acre** "Sample Costs to Establish a Vineyard and Produce Winegrapes," U.C. Cooperative Extension, 2012, https://coststudyfiles.ucdavis.edu/uploads/cs_public/c7/0b/c70be1f3-aa8c-40c5-b7ac-5fade99a2e2b/grapewinevn2012.pdf.

68 **Paso Robles** Tom Philpott, "Harvard Is Buying Up Vineyards in Drought-Ridden California Wine Country," *Mother Jones*, January 31, 2015, http://www.motherjones.com/food/2015/01/wine-water-harvards-move-california-farmland.

69 **changed directions** Michael McDonald and Sabrina Willmer, "Harvard Said to Seek Sale of Slumping Natural Resources Assets," *Bloomberg Quint*,

November 11, 2016, https://www.bloombergquint.com/business/2016/11/10/harvard-said-to-seek-sale-of-slumping-natural-resources-assets.

69 **more sales** Juliet Chung, Dawn Lim, and Lucy Craymer, "The Harvard Yard Sale: Private Equity, Real Estate and New Zealand Dairy Farms," *Wall Street Journal*, June 6, 2017, https://www.wsj.com/articles/the-harvard-yard-sale-private-equity-real-estate-and-new-zealand-dairy-farms-1496741403.

69 **"critically overdrafted basins"** https://web.archive.org/web/20180128082124/http://www.water.ca.gov/groundwater/sgm/pdfs/COD_BasinsTable.pdf.

70 **Land with access to surface water** "Agricultural Trends in Value for Kern County," Alliance Ag, https://www.allianceappr.com/trends-in-value-1.

CHAPTER 4: EMPIRE OF DIRT

73 **90 percent of the state's** "Crop and Land Use: Statewide Data," Soil and Land Use, Iowa State University, https://www.extension.iastate.edu/soils/crop-and-land-use-statewide-data.

73 **hundreds of species** Mark Muller, "Iowa Prairie Plants," University of Iowa Library, http://uipress.lib.uiowa.edu/ppi/preface.php; and Jill Haukos, "The Tallgrass Prairie," University of Kansas, https://keep.konza.k-state.edu/prairieecology/TallgrassPrairieEcology.pdf.

73 **played an active role** National Park Service, "Fire and Grazing in the Prairie," accessed March 8, 2019, https://www.nps.gov/tapr/learn/nature/fire-and-grazing-in-the-prairie.htm.

74 **Iowa's tallgrass prairie** Neil Smith, "Tallgrass Prairie," U.S. Fish and Wildlife Service, accessed March 2, 2019, https://www.fws.gov/refuge/Neal_Smith/wildlife_and_habitat/tallgrass_prairie.html.

74 **surrendered fully one half** Laura Miller, "Cost of Soil Erosion in Iowa, Not a Pretty Picture," *Leopold Letter*, Fall 2015, Leopold Center for Sustainable Agriculture, Iowa State University. See also "Visualizing Iowa's Topsoil Loss," Boing Boing, May 4, 2011, https://boingboing.net/2011/05/04/visualizing-iowas-to.html.

75 **Iowa farmers churned out** Mark Weinraub, "USDA Expects Record Soy Production, Corn Yields," Reuters, August 10, 2018, accessed March 3, 2019, https://www.reuters.com/article/us-usda-crops/usda-expects-record-soy-production-corn-yields-idUSKBN1KV22Z.

75 **Iowa's hog population** USDA National Agricultural Statistics Service, "Iowa Ag News—Hogs and Pigs," September 27, 2018, https://www.nass.usda.gov/Statistics_by_State/Iowa/Publications/Livestock_Report/2018/IA-Hogs-09-18.pdf.

75 **center of the Corn Belt** Timothy R. Green, Holm Kipka, Olaf David, and Gregory S. McMaster, "Where Is the USA Corn Belt, and How Is It Changing?," USDA Agricultural Research Service 2017, accessed January 13, 2019, https://digitalcommons.unl.edu/cgi/viewcontent.cgi?article=2848&context=usdaarsfacpub. Corn: "World of Corn 2018," accessed March 17, 2019, http://www.worldofcorn.com. Soybeans: "World Soybean Production," March 2018, Soy Bean Meal Info Center, accessed March 17, 2019, https://www.soymeal.org/soy-meal-articles/world-soybean-production.

76 **record-breaking U.S. meat habit** OECD, "Meat Consumption," accessed May 2, 2019, https://data.oecd.org/agroutput/meat-consumption.htm.

76 **Meat, dairy, and eggs** "How America's Diet Has Changed over the Decades," Pew Research, December 13, 2016, http://www.pewresearch.org/fact-tank/2016/12/13/whats-on-your-table-how-americas-diet-has-changed-over-the-decades.

76 **Soybean oil alone** "Management of Pioneer® brand Plenish® High Oleic Soybean Management," Pioneer, accessed March 2, 2019, https://www.pioneer.com/home/site/us/agronomy/library/plenish-high-oleic-soybean-mgmt; and "Soybeans and the U.S. Food Supply," Soy Nutrition Institute, May 4, 2016, https://thesoynutritioninstitute.com/soybeans-and-the-u-s-food-supply.

76 **corn-derived sugars** USDA Economic Research Service, "Sugar and Sweeteners Yearbook Tables," https://www.ers.usda.gov/data-products/sugar-and-sweeteners-yearbook-tables.aspx.

77 **corn crop is transformed** Energy Information Agency, "Almost All U.S. Gasoline Is Blended with 10% Ethanol," May 4, 2016, https://www.eia.gov/todayinenergy/detail.php?id=26092.

77 **Soy-derived biodiesel** "Biodiesel," United Soybean Board, https://unitedsoybean.org/media-center/issue-briefs/biodiesel.

77 **More than 50 percent of U.S. soybeans** "Soybean Markets Vital as Exports Top 60% of U.S. Production," Farm Futures, https://www.farmprogress.com/story-soybean-markets-vital-exports-top-60-production-8-131125;

"World of Corn 2018," http://www.worldofcorn.com; and "Pork Facts," National Pork Producers Council, http://nppc.org/pork-facts.

77 **middle-class and wealthy consumers** Environmental Working Group, "Feeding the World," October 5, 2016, https://www.ewg.org/research/feeding-the-world.

78 **In 1940, Iowa supported** U.S. Census Bureau, *Agriculture Census*, 1950, https://www2.census.gov/library/publications/decennial/1950/agriculture-volume-2/21895591v2ch02.pdf.

78 **total number of farms** USDA National Agricultural Statistics Service, "Census of Agriculture: State Profile—Iowa," 2012, https://www.nass.usda.gov/Publications/AgCensus/2012/Online_Resources/County_Profiles/Iowa/cp99019.pdf.

78 **Iowa lost nearly** USDA National Agricultural Statistics Service, "2018 State Agriculture Overview: Iowa," https://www.nass.usda.gov/Quick_Stats/Ag_Overview/stateOverview.php?state=IOWA.

79 **median Iowa farm surged** James M. MacDonald, Penni Korb, and Robert A. Hoppe, *Farm Size and the Organization of US Crop Farming*, August 2013, USDA Economic Research Service, https://www.ers.usda.gov/webdocs/publications/45108/39359_err152.pdf.

79 **U.S. agricultural output rose** Sun Ling Wang, Paul Heisey, David Schimmelpfennig, and V. Eldon Ball, "Agricultural Productivity Growth in the United States: Measurement, Trends, and Drivers," Economic Research Service, Paper No. ERR-189, 2015.

79 **exist on a technology treadmill** Daryll E. Ray, Daniel G. De La Torre Ugarte, and Kelly H. Tiller, "Rethinking U.S. Agricultural Policy: Changing Course to Secure Farmer Livelihoods Worldwide," University of Tennessee, Agricultural Policy Analysis Center, https://www.agpolicy.org/blueprint/APACReport8-20-03WITHCOVER.pdf.

79 **those external factors** Marco Lagi, Yavni Bar-Yam, Karla Z. Bertrand, and Yaneer Bar-Yam, "The Food Crises: A Quantitative Model of Food Prices Including Speculators and Ethanol Conversion," accessed April 1, 2019, https://arxiv.org/abs/1109.4859. See also "Food Commodities Speculation and Food Price Crises: Regulation to Reduce the Risks of Price Volatility," briefing note by Olivier De Schutter, United Nations special rapporteur on the right to food, September 2010, https://www2.ohchr.org/english/issues/food/docs/Briefing_Note_02_September_2010_EN.pdf.

80 **"Agricultural returns tend"** Chad Hart, "Ag Cycles," Iowa State University Extension and Outreach, https://www.extension.iastate.edu/agdm/info/agcycles/hart.pdf.

80 **corn and soybean costs** USDA Economic Research Service, "Commodity Costs and Returns," February 5, 2019, https://www.ers.usda.gov/data-products/commodity-costs-and-returns.

81 **received nearly $149 billion** Environmental Working Group Farm Subsidy Database, Iowa Farm Subsidy Information, https://farm.ewg.org/region.php?fips=19000&statename=Iowa.

82 **John Deere alone** Jennifer Reibel, "Manufacturer Consolidation Reshaping the Farm Equipment Market," *Farm Equipment*, August 29, 2018, https://www.farm-equipment.com/articles/15962-manufacturer-consolidation-reshaping-the-farm-equipment-marketplace.

82 **new combine** See the John Deer website, https://configure.deere.com/cbyo/#/en_us/products.

82 **combine with an engine rated** "Grain Harvesting Equipment and Labor in Iowa," Iowa State University, https://www.extension.iastate.edu/agdm/crops/html/a3-16.html.

83 **"right to repair" bills** Margaret Sessa-Hawkins, "In Fight over the Right to Repair Equipment, Farmers Are Outspent 28 to 1," June 6, 2017, Maplight, https://maplight.org/story/in-fight-over-the-right-to-repair-equipment-farmers-are-outspent-28-to-1.

83 **Deere's website** "Equipment Loans," John Deere, accessed May 4, 2019, https://www.deere.com/en/finance/financing/equipment-loans.

83 **Deere's financial-services division** John Deere Financial Fact Book, 2018, accessed March 14, 2019, https://s22.q4cdn.com/253594569/files/doc_downloads/books/2019/JDF-Fact-Book_2018-(1).pdf; and Jesse Newman and Bob Tita, "America's Farmers Turn to Bank of John Deere," *Wall Street Journal*, July 18, 2017, https://www.wsj.com/articles/americas-farmers-turn-to-bank-of-john-deere-1500398960.

84 **U.S. fertilizer market** "United States Fertilizers Market—Size, Share, Analysis (2017–2022)," Mordor Intelligence, https://www.mordorintelligence.com/industry-reports/united-states-fertilizers-market.

84 **U.S. potash and phosphate markets** "View Fact Book," Nutrien, https://www.nutrien.com/sites/default/files/uploads/2018-01/Nutrien%20Fact%20Book%202018_1.pdf.

84 **Mosaic's position in the global** Seth Goldstein, "Mosaic Reduces Phosphate Production to Support Prices; Maintaining $37 Fair Value Estimate," Morningstar, accessed March 17, 2019.

85 **extraordinary market concentration** Trevor Brown, "CF-OCI Merger Cancelled," *Ammonia Industry*, May 23, 2016, https://ammoniaindustry.com/cf-oci-merger-cancelled.

85 **Monsanto dropped $240 million** "Monsanto to Buy Asgrow Agronomics Unit," *New York Times*, September 25, 1996, https://www.nytimes.com/1996/09/25/business/monsanto-to-buy-asgrow-agronomics-unit.html.

85 **another $319 million** George Gunset, "Monsanto to Buy up to 40 percent of Dekalb Genetics," February 2, 1996, https://www.chicagotribune.com/news/ct-xpm-1996-02-02-9602020051-story.html.

85 **Pioneer Hi-Bred International** Steven Lipin, Scott Kilman, and Susan Warren, "DuPont Agrees to Purchase of Seed Firm for $7.7 Billion," *Wall Street Journal*, March 15, 1999, https://www.wsj.com/articles/SB921268716949898331.

85 **"Big Six"** USDA Economic Research Service, James M. MacDonald, "Mergers and Competition in Seed and Agricultural Chemical Markets," https://www.ers.usda.gov/amber-waves/2017/april/mergers-and-competition-in-seed-and-agricultural-chemical-markets.

86 **$68 billion global seed market** "Global $99 Bn Seed Market 2017–2022 by Type, Trait, Crop Type an Regions—Research and Markets," *Business Wire*, https://www.businesswire.com/news/home/20171117005380/en/Global-99-Bn-Seed-Market-2017-2022-Type.

86 **Big Three loom even larger** Sonja Begemann, "What to Watch as Mergers Near the Finish Line," AgWeb, August 2, 2017, https://www.agweb.com/article/what-to-watch-as-mergers-near-the-finish-line-naa-sonja-begemann.

86 **Pat Mooney, a longtime industry watcher** Pat Mooney, "Blocking the Chain," ETC Group, October 15, 2018, https://www.etcgroup.org/content/blocking-chain.

87 **delivered consistently low crop yields** Bruno Basso, Guanyuan Shuai, Jinshui Zhang, and G. Philip Robertson, "Yield Stability Analysis Reveals Sources of Large-Scale Nitrogen Loss from the US Midwest," *Scientific Reports*, April 8, 2019.

88 **company was claiming** Climate Corporation press release, June 24, 2015, https://climate.com/newsroom/digital-agronomic-services-platform/12.

88 **company's crown jewel** "Monsanto (MON) Earnings Report: Q3 2016 Conference Call Transcript," June 29, 2016, https://www.thestreet.com/story/13625066/1/monsanto-mon-earnings-report-q3-2016-conference-call-transcript.html.

88 **Climate Fieldview as the hub** Louisa Burwood-Taylor, "Monsanto's Fraley on Staying Ahead in Agtech Innovation," March 17, 2016, https://agfundernews.com/monsantos-fraley-on-staying-ahead-in-agtech-innovation5552.html.

89 **can "export prescriptions"** "John Deere Operations Center," Climate Field View, https://climate.com/friends/john-deere-operations-center.

89 **Corteva Agriscience has a deal with Planet** Lee Smith, "DowDuPont's Granular Taps Planet for Multi-Year Deal for Agriculture," Planet, March 20, 2018, https://www.planet.com/pulse/dowdupont-granular-tap-planet.

89 **deal with John Deere** "Deere and Granular Introduce New Farm Management Tool," John Deere, https://www.deere.com/en/our-company/news-and-announcements/news-releases/2018/agriculture/2018sep14-profit-maps-farm-management-tool.

89 **complete with a satellite offering** "Syngenta Acquires Satellite Imagery Provider FarmShots," PrecisionAg," February 14, 2018, https://www.precisionag.com/in-field-technologies/imagery/syngenta-acquires-satellite-imagery-provider-farmshots.

90 **global grain-trading market** Sophia Murphy, *Cereal Secrets: The World's Largest Grain Traders and Global Agriculture*, Oxfam International, August 2012, p. 78, https://www-cdn.oxfam.org/s3fs-public/file_attachments/rr-cereal-secrets-grain-traders-agriculture-30082012-en_4.pdf.

90 **process 85 percent of U.S.-grown soybeans** "The Dynamic State of Agriculture and Food: Possibilities for Rural Development?," Statement of Mary Hendrickson, Ph.D., University of Missouri, at the Farm Credit Administration Symposium on Consolidation in the Farm Credit System, McLean, Virginia, February 19, 2014.

90 **Bunge is king in Brazil** Tom Polansek, "ADM Approach to Bunge Marks Potential U-turn on Oilseed Strategy," Reuters, January 22, 2018, https://www.reuters.com/article/us-bunge-m-a-archer-daniels/adm-approach-to-bunge-marks-potential-u-turn-on-oilseed-strategy-idUSKBN1FC000.

90 **ramps up their competition** Tatiana Freitas and Megan Durisin, "The Great Corn Clash Is Coming as U.S., Brazil Farmers Face Off," Bloomberg News,

July 30, 2017, https://www.bloomberg.com/news/articles/2017-07-30/the-great-corn-clash-is-coming-as-u-s-brazil-farmers-face-off.

90 **paved the way** Tom Philpott, "How Cash and Corporate Pressure Pushed Ethanol to the Fore," *Grist*, December 7, 2006, https://grist.org/article/adml.

90 **15 billion gallons of ethanol** "EPA Finalizes Renewable Fuel Standard for 2019," EIA, December 6, 2018, https://www.eia.gov/todayinenergy/detail.php?id=37712.

91 **claims are fiercely contested** See, for example, Timothy D. Searchinger, Stefan Wirsenius, Tim Beringer, and Patrice Dumas, "Assessing the Efficiency of Changes in Land Use for Mitigating Climate Change," *Nature* 564, no. 7735 (2018): 249, DOI: 10.1038/s41586-018-0757-z. Searchinger, a Princeton researcher, and his team calculate that corn-based ethanol emits more greenhouse gas per gallon than petroleum gasoline.

91 **biggest sources of profit for Cargill** "Cargill's Bets on Animal Protein Pay Off as Profits Hit Record," *Financial Times*, July 12, 2018, https://www.ft.com/content/df258fec-8541-11e8-a29d-73e3d454535d.

91 **Brazilian giant JBS** "Brazil's JBS Buys Swift Foods for $1.4 Bln," MarketWatch," May 29, 2007, accessed March 21, 2019, https://www.marketwatch.com/story/brazils-jbs-buys-swift-foods-for-14-bln.

91 **account for 85 percent** USDA Grain Inspection, Packers, and Stockyard Administration, "2016 Packers and Stockyards Program Annual Report," accessed March 20, 2019, https://www.gipsa.usda.gov/psp/publication/ar/2016_psp_annual_report.pdf.

92 **massive amounts of meat** USDA Economic Research Service, "Per Capita Red Meat and Poultry Disappearance: Insights," accessed March 22, 2019, https://www.ers.usda.gov/amber-waves/2018/june/per-capita-red-meat-and-poultry-disappearance-insights-into-its-steady-growth.

92 **more meat per capita** OECD, "Meat Consumption."

92 **meat consumption is on the rise** Johnny Wood, "How Our Growing Appetite for Meat Is Harming the Planet," accessed March 22, 2019, World Economic Forum, August 21, 2018, https://www.weforum.org/agenda/2018/08/global-appetite-for-meat-is-growing.

92 **interview with journalist Christopher Leonard** Leonard and Tyson quoted in Tom Philpott, "Don Tyson Details Plans to Export the US Meat Model to Global," *Grist*, November 12, 2008, https://grist.org/article/meat-wagon-all-the-worlds-a-cafo.

92 **U.S. beef exports jumped** "Statistics," U.S. Meat Export Federation, https://www.usmef.org/news-statistics/statistics.

92 **exports had reached** "U.S. Broiler Exports Quantity and Share of Production," National Chicken Council, December 18, 2018, https://www.nationalchickencouncil.org/about-the-industry/statistics/u-s-broiler-exports-quantity-and-share-of-production.

92 **U.S. meat exports surged** USDA Economic Research Service, "Per Capita Red Meat and Poultry Disappearance: Insights."

92 **more than 25 percent of U.S. pork** "U.S. Pork Exports Set New Volume Records in 2017," Pork Checkoff, February 7, 2018, https://www.pork.org/news/u-s-pork-exports-set-new-volume-records-2017; and "FAQ," U.S. Meat Export Federation, https://www.usmef.org/about-usmef/faq.

93 **main difference is feed costs** Fred Gale, Daniel Marti, and Dinghuan Hu, "China's Volatile Pork Industry," February 2012, USDA Economic Research Service, https://www.ers.usda.gov/webdocs/publications/37433/13744_ldpm21101_1_.pdf?v=0.

93 **poultry industry was in the midst** "Infographic: 7 New Broiler Plants to Begin Operations," *WattAgNet*, October 21, 2018, https://www.wattagnet.com/articles/35638-infographic-7-new-broiler-plants-to-begin-operations.

93 **production will likely expand 3 percent annually** "Can Americans Eat More Chicken? U.S. Industry Bets on Growth," *Bloomberg News*, February 7, 2018, https://www.bloomberg.com/news/articles/2018-02-07/can-americans-eat-more-chicken-industry-bets-windfall-on-growth.

93 **massive new packing plant** "Pork Industry Expands, But at Controlled Pace," *National Hog Farmer*, September 29, 2017, https://www.nationalhogfarmer.com/marketing/pork-industry-expands-controlled-pace.

94 **average hourly wage** Bureau of Labor Statistics, "Slaughterers and Meat Packers," https://www.bls.gov/oes/2017/may/oes513023.htm.

94 **repetitive-motion conditions** "Working 'the Chain,' Slaughterhouse Workers Face Lifelong Injuries," National Public Radio, August 11, 2016, https://www.npr.org/sections/thesalt/2016/08/11/489468205/working-the-chain-slaughterhouse-workers-face-lifelong-injuries.

94 **speed at which they work** USDA Food Safety and Inspection Service, "Evaluation of HACCP Inspection Models Project (HIMP) for Market Hogs," https://www.fsis.usda.gov/wps/wcm/connect/f7be3e74-552f-4239-ac4c-59a024fd0ec2/Evaluation-HIMP-Market-Hogs.pdf?MOD=AJPERES.

94 **refugees fleeing political violence** "Tighter Refugee Rules Seen Squeezing Meat Companies," *Wall Street Journal*, January 28, 2017, https://www.wsj.com/articles/tighter-refugee-rules-seen-squeezing-meat-companies-1485617876.

95 **environmental scientist Jonathan Foley** Jonathan Foley, "It's Time to Rethink America's Corn System," Ensia, March 5, 2013, https://ensia.com/voices/its-time-to-rethink-americas-corn-system.

95 **60 percent of the calories** Eurídice Martínez Steele, "Ultra-Processed Foods and Added Sugars in the US Diet: Evidence from a Nationally Representative Cross-Sectional Study," *BMJ* 6, no. 3 (2016): e009892.

96 **one or more chronic diseases** Magdalena M. Wilson, "American Diet Quality: Where It Is, Where It Is Heading, and What It Could Be," *Journal of the Academy of Nutrition and Dietetics* 116, no. 2 (2015): 302–10.el.

96 **Farm bankruptcies in the region** "Farm Bankruptcies in 2018: The Truth is Out There," Farm Bureau, February 12, 2019, https://www.fb.org/market-intel/farm-bankruptcies-in-2018-the-truth-is-out-there.

96 **Farm debt reached** Jesse Newman and Jacob Bunge, "US Farmers, Who Once Fed the World, Are Overtaken by New Powers," *Wall Street Journal*, April 20, 2017, https://www.wsj.com/articles/u-s-farmers-who-once-fed-the-world-overtaken-by-new-world-powers-1492700574.

98 **sped it up in another vast region** Yue Dou, Bicudo da Silva, Ramon Felipe Hongbo, and Jianguo Liu Yang, "Spillover Effect Offsets the Conservation Effort in the Amazon," *Journal of Geographical Sciences*, November 2018.

98 **supported human populations** "Ecosystem Profile: Cerrado Biodiversity Hotspot Extended," February 2017, Critical Ecosystem Partnership Fund, https://www.cepf.net/resources/documents/cerrado-ecosystem-profile-summary-revised-2017.

99 **dozens of indigenous groups** "The Brazilian Government's Land War Against Rebel Slave Descendants," Mongabay, October 29, 2018, https://news.mongabay.com/2018/10/the-brazilian-governments-land-war-against-rebel-slave-descendants.

99 **251 species of mammals** Myan Lahsen, Mercedes M. C. Bustamante, and Eloi L. Dalla-Nora, "Undervaluing and Overexploiting the Brazilian Cerrado at Our Peril," *Environment: Science and Policy for Sustainable Development* 58, no. 6 (2016): 4–15.

99 **"one of the richest places"** "Brazil's Threatened Cerrado Gets a Protection Plan," World Wildlife Fund, September 22, 2010, http://wwf.panda.org/wwf_news/?195072/Brazils-threatened-Cerrado-gets-a-protection-plan.

99 **"inverted forest"** Myan, Bustamante, and Dalla-Nora, "Undervaluing and Overexploiting the Brazilian Cerrado."

99 **financiers and large landholders** "Brazil Senator Maggi Joins Forbes Billionaire List," April 10, 2014, https://www.forbes.com/sites/kenrapoza/2014/04/10/for-brazil-farmer-gift-of-the-maggi-worth-a-billion-bucks; "Soros-Backed Grain Company Eyes Trump Trade in Brazil," *Financial Times*, February 6, 2017, https://www.ft.com/content/6da964a8-ebd5-11e6-930f-061b01e23655; and Fred Pearce, "The Cerrado: Brazil's Other Biodiverse Region Loses Ground," *Yale Environment 360*, April 14, 2011, https://e360.yale.edu/features/the_cerrado_brazils_other_biodiversity_hotspot_loses_ground.

99 **holdings have embroiled** Cerrado: US Investment Spurs Land Theft and Deforestation in Brazil," March 29, 2018, https://www.grain.org/bulletin_board/entries/5926-cerrado-us-investment-spurs-land-theft-and-deforestation-in-brazil.

100 **land has succumbed to agriculture** Pearce, "The Cerrado."

100 **extolled the plowing of the Cerrado** Borlaug quoted in Tsuioshi Yamada "The Cerrado of Brazil: A Success Story of Production on Acid Soils, Soil Science, and Plant Nutrition," 2005, https://doi.org/10.1111/j.1747-0765.2005.tb00076.x.

100 **carbon emissions from expanding cropland** Praveen Noojipady et al., "Forest Carbon Emissions from Cropland Expansion in the Brazilian Cerrado Biome," *Environmental Research Letters*, February 2, 2017.

100 **year's tailpipe carbon emissions** According to Noojipady et al., expansion of farmland in the Cerrado emits on average 16.3 million metric tons of carbon per year. An EPA calculator—at epa.gov/energy/greenhouse-gas-equivalencies-calculator—finds that's equal to emissions from 12.7 million passenger cars driven per year. There were 6.4 million passenger cars registered in Los Angeles County in 2017. http://www.laalmanac.com/transport/tr02.php.

100 **carbon from the soil to the atmosphere** W. J. Parton, M. P. Gutmann, E. R. Merchant, et al., "Measuring and Mitigating Agricultural Greenhouse

Gas Production in the US Great Plains, 1870–2000," *Proceedings of the National Academy of Sciences* 112, no. 34 (2015): E4681–88, DOI: 10.1073/pnas.1416499112.

101 **Brazil "emerged as the largest"** USDA Economic Research Service, "Brazil's Corn Industry and the Effect on the Seasonal Prices," https://www.ers.usda.gov/webdocs/publications/35806/59643_aes93.pdf?v=0.

101 **Brazil overtook the United States** USDA Agricultural Marketing Service, "The Impact of Infrastructure and Transportation Costs on U.S. Soybean Prices," https://www.ams.usda.gov/sites/default/files/media/SoybeanMarketShare19922017.pdf.

101 **Brazil had 53 percent** Karl Plume, "Protein Plight: Brazil Steals U.S. Soybean Share in China," Reuters, January 25, 2018, https://www.reuters.com/article/us-usa-soybeans-protein-insight/protein-plight-brazil-steals-u-s-soybean-share-in-china-idUSKBN1FE0FM.

102 **Hugh Grant, tells the story** Jacob Bunge, "Life After Monsanto Sale," *Wall Street Journal*, June 21, 2018, https://www.wsj.com/articles/life-after-the-monsanto-sale-1529586001, https://aflcio.org/paywatch. Also see Salary.com, https://www1.salary.com/Hugh-Grant-Salary-Bonus-Stock-Options-for-MONSANTO-CO.html.

CHAPTER 5: FAILING UPWARD

105 **Monsanto had known about** "Anniston, Alabama: Monsanto Knew About PCB Toxicity for Decades," Environmental Working Group, March 7, 2009, https://www.ewg.org/research/anniston-alabama/monsanto-knew-about-pcb-toxicity-decades.

105 **agreed to pay a $390 million** "Monsanto Contributes to Global Settlement of Anniston, Alabama, PCB Litigation," Monsanto press release, August 20, 2003, https://monsanto.com/news-releases/monsanto-contributes-to-global-settlement-of-anniston-alabama-pcb-litigation.

105 **Corn Belt** Gale Peterson, "The Discovery and Development of 2,4-D," *Agricultural History* 41, no. 3 (1967): 243–54, www.jstor.org/stable/3740338.

106 **$270 million** Peterson, "Discovery and Development of 2,4-D."

106 ***Silent Spring*** Rachel Carson, *Silent Spring* (New York: Houghton Mifflin, 1962).

106 **factor of three** Jorge Fernandez-Cornejo, Richard Nehring, Craig Osteen, Seth Wechsler, Andrew Martin, and Alex Vialou, "Pesticide Use in U.S. Agriculture: 21 Selected Crops, 1960–2008," EIB-124, USDA Economic Research Service, May 2014, https://www.ers.usda.gov/webdocs/publications/43854/46734_eib124.pdf.

106 **Agent Orange** Ralph Blumenthal, "Veterans Accept $180 Million Pact on Agent Orange," *New York Times*, May 8, 1984, https://www.nytimes.com/1984/05/08/nyregion/veterans-accept-180-million-pact-on-agent-orange.html.

106 **U.S. soldiers** Clyde Haberman, "Agent Orange's Long Legacy, for Vietnam and Veterans," *New York Times*, May 11, 2014, https://www.nytimes.com/2014/05/12/us/agent-oranges-long-legacy-for-vietnam-and-veterans.html.

106 **Vietnamese civilians** Jessica King, "U.S. in First Effort to Clean Up Agent Orange in Vietnam," CNN.com, August 10, 2012, https://www.cnn.com/2012/08/10/world/asia/vietnam-us-agent-orange/index.html.

107 **amino acids** "Glyphosate—an Overview," ScienceDirect Topics, https://www.sciencedirect.com/topics/neuroscience/glyphosate.

107 **Roundup gained traction** Daniel Charles, *Lords of the Harvest: Biotech, Big Money, and the Future of Food* (New York: Basic Books, 2001), 63.

107 ***Diamond v. Chakrabarty*** 447 U.S. 303 (1980). No. 79-139, United States Supreme Court, June 16, 1980, https://caselaw.findlaw.com/us-supreme-court/447/303.html.

107 **Daniel Charles** Charles, *Lords of the Harvest*, 60–62.

109 **shelved plans for Roundup Ready wheat** "Monsanto Puts Hold on Roundup Ready Wheat," *No-Till Farmer*, July 1, 2004, https://www.no-tillfarmer.com/articles/3079-monsanto-puts-hold-on-roundup-ready-wheat.

109 **feeding the world** Monsanto Global, http://www.monsantoglobal.com/global/ph/whoweare/Pages/default.aspx.

109 **lead to resistant weeds** USDA Animal and Plant Health Inspection Service, "Petition for Determination of Non-Regulated Status: Soybeans with a Roundup Ready™ Gene, Monsanto #93-089U," September 14, 1993, http://www.aphis.usda.gov/brs/aphisdocs/93_25801p.pdf.

109 **stave off the potent chemical** Laura Bradshaw, Stephen Padgette, Steven Kimball, and Barbara Wells, "Perspectives on Glyphosate Resistance," *Weed Technology* 11, no. 1 (1997): 189–98, https://www.jstor.org/stable/3988252?seq=1.

110 **Farmers controlled weeds** Charles Benbrook, "Impacts of Genetically Engineered Crops on Pesticide Use in the U.S.—the First Sixteen Years," *Environmental Sciences Europe* 24, no. 24 (2012), https://doi.org/10.1186/2190-4715-24-24.

110 **low prices due to crop overproduction** Richard Brock, "Repeat of Late 1990s for Corn and Soybeans?," *Western Farm Press*, October 5, 2017, https://www.farmprogress.com/corn/brock-repeat-late-1990s-corn-and-soybeans.

111 **looming resistance crisis** Bradshaw et al., "Perspectives on Glyphosate Resistance."

111 **not just Roundup** Bob Hartzler, "No Benefit in Rotating Glyphosate," Iowa State University, December 17, 2004, http://extension.agron.iastate.edu/weeds/mgmt/2004/twoforone.shtml.

111 **glyphosate-resistant waterhemp** Jackie Pucci, "The War Against Weeds Evolves in 2018," *CropLife*, March 20, 2018, https://www.croplife.com/crop-inputs/the-war-against-weeds-evolves-in-2018.

112 **eight weed species** Vijay Nandula, Krishna Reddy, Stephen Duke, and Daniel Poston, "Glyphosate-Resistant Weeds: Current Status and Future Outlook," *Outlooks on Pest Management* 16, no. 4 (2005): 183–87, https://www.ars.usda.gov/ARSUserFiles/64022000/Publications/Reddy/Nandula-GRW12.pdf.

112 **glyphosate was so widely used** Feng-chih Chang, Matt Simcik, and Paul Capel, "Occurrence and Fate of the Herbicide Glyphosate and Its Degradate Aminomethylphosphonic Acid in the Atmosphere," *Environmental Toxicology and Chemistry* 30, no. 3 (2011): 548–55, DOI: 10.1002/etc.431. See also USGS, "Widely Used Herbicide Commonly Found in Rain and Streams in the Mississippi River Basin," technical announcement, August 29, 2011, https://archive.usgs.gov/archive/sites/www.usgs.gov/newsroom/article.asp-ID=2909.html.

113 **opportunity to sell more herbicides** Jack Kaskey, "Attack of the Superweed," *Bloomberg Businessweek*, September 8, 2011, https://www.bloomberg.com/news/articles/2011-09-08/attack-of-the-superweed.

113 **dicamba-Roundup cocktail** USDA Animal and Plant Health Inspection Service, "Petition for the Determination of Nonregulated Status for Dicamba-Tolerant Soybean MON 87708," Monsanto, July 6, 2010, p. 594, https://www.aphis.usda.gov/brs/aphisdocs/10_18801p.pdf.

113 **2,4-D resistant weeds** USDA Animal and Plant Health Inspection Service, "Petition for Determination of Nonregulated Status for Herbicide Tolerant

DAS-68416-4 Soybean," Dow AgroSciences, November 17, 2010, pp. 5, 174, https://www.aphis.usda.gov/brs/aphisdocs/09_34901p.pdf.

114 **shredded that argument** David Mortensen, J. Franklin Egan, Bruce D. Maxwell, Matthew R. Ryan, and Richard G. Smith, "Navigating a Critical Juncture for Sustainable Weed Management," *BioScience* 62, no. 1 (2012): 75–84.

114 **weeds that can resist dicamba, 2,4-D, and glyphosate** "Palmer Amaranth Resistance to 2,4-D and Dicamba Confirmed in Kansas," Kansas State University Research and Extension, March 1, 2019, https://webapp.agron.ksu.edu/agr_social/m_eu_article.throck?article_id=2110&eu_id=322.

114 **dicamba is highly volatile** Public Participation for Dicamba: New Use on Herbicide-Tolerant Cotton and Soybean, "Comment submitted by Jerry Philip W. Miller, PhD, Vice President, Global Regulatory and Government Affairs, Monsanto Company," https://www.regulations.gov/document?D=EPA-HQ-OPP-2016-0187-0858.

115 **43 percent of the U.S. soybean crop** USDA Animal and Plant Health Inspection Service, "Petition for the Determination of Nonregulated Status of Dicamba-Tolerant Soybean MON 87708."

115 **conquered 43 percent** USDA Economic Research Service, "The Use of Genetically Engineered Dicamba-Tolerant Soybean Seeds Has Increased Quickly, Benefiting Adopters but Damaging Crops in Some Fields," October 1, 2019.

116 ***Bill Nye the Science Guy*** Bill Nye, *Undeniable: Evolution and the Science of Creation* (New York: St. Martin's Press, 2014).

117 **backstage on Bill Maher's HBO show** *Real Time with Bill Maher*, "Backstage with Bill Nye the Science Guy," February 25, 2015, HBO, https://www.youtube.com/watch?v=81FAqQ1qSiQ&feature=youtu.be.

117 **"More Food, Less Hype"** *Bill Nye Saves the World*, season 1, Netflix, https://www.youtube.com/watch?v=81FAqQ1qSiQ&feature=youtu.be.

122 **return less than a third that size** Derived from the https://dqydj.com/stock-return-calculator-dividend-reinvestment-drip website.

122 **"You can only squeeze"** Jonas Oxgaard email to author.

CHAPTER 6: GULLY WASHERS

124 **photosynthetic activity** Luis Guanter et al., "Global and Time-Resolved Monitoring of Crop Photosynthesis with Chlorophyll Fluorescence,"

Proceedings of the National Academy of Sciences 111, no. 14 (2014): E1327–33, https://doi.org/10.1073/pnas.1320008111.

125 **bomb cyclone** Mitch Smith, Jack Healy, and Timothy Williams, "'It's Probably Over for Us': Record Flooding Pummels Midwest When Farmers Can Least Afford It," *New York Times*, March 18, 2019, https://www.nytimes.com/2019/03/18/us/nebraska-floods.html.

129 **5-ton assumption** Dennis Keeney and Rick Cruse, "Policy and Government Programs in Soil and Water Conservation," in *Advances in Soil and Water Conservation*, ed. F. J. Pierce and W. W. Frye (Chelsea, MI: Ann Arbor Press, 1998): 185–94.

129 **around 0.5 tons per acre** Earl Alexander, "Rates of Soil Formation: Implications for Soil-loss Tolerance," *Soil Science* 145, no. 1 (1988): 37–45. See also "Erosion's Long Destructive Train," Environmental Working Group, https://www.ewg.org/losingground/report/erosions-long-destructive-train/3.html.

129 **ephemeral gullies** Lee Gordon, Sean Bennett, Carlos Alonso, and Ronald Bingner, "Modeling Long-Term Soil Losses on Agricultural Fields Due to Ephemeral Gully Erosion," *Journal of Soil and Water Conservation* 63, no. 4 (2008): 173–81. This paper puts the long-term average of soil lost to gully erosion at 2.23 tons to 4.91 tons per acre per year. Cruse uses 3 tons per acre as his working assumption. In 1996, the USDA's Natural Resources Conservation Service estimated Iowa's soil losses to ephemeral gullies at 3 tons per acre, and those of neighboring Illinois, under a similar mat of corn and soybeans, at 5 tons per acre. See *America's Private Land, A Geography of Hope*, part 2, p. 39, Natural Resources Conservation Service, https://www.nrcs.usda.gov/wps/portal/nrcs/detail/national/technical/nra/rca/?cid=nrcs143_014212.

130 **corn yields plunged** Bradley Rippey, "The U.S. Drought of 2012," *Weather and Climate Extremes* 10, part A (December 2015), 57–64, https://www.sciencedirect.com/science/article/pii/S2212094715300360.

131 **precipitation patterns** U.S. Global Change Research Program, "Midwest," ch. 21 in *Impacts, Risks, and Adaptation in the United States: Fourth National Climate Assessment, Vol. 2*, ed. K.L.M. Lewis, D. R. Reidmiller, C. W. Avery, et al. (Washington, D.C.: Government Printing Office, 2018), https://nca2018.globalchange.gov/chapter/21, DOI: 10.7930/NCA4.2018.

132 **spring 2019's heavy storms** U.S. Global Change Research Program, *Impacts, Risks, and Adaptation in the United States*, "Figure 2.6: Observed and Projected Change in Heavy Precipitation," https://nca2018.globalchange.gov/chapter/2/#fig-2-6.

132 **causing all the mayhem** Iowa Department of Agriculture and Land Stewardship, Monthly Weather Reports: March 2019: 2.1 inches; April 2019: 2.96 inches; May 2019: 7.93 inches, https://iowaagriculture.gov/climatology-bureau/iowa-historic-weather-reports.

132 **141 years of weather recordkeeping** Rod Swoboda, "2013 Is Wettest Spring on Record in Iowa," *Farm Progress*, May 29, 2013, https://www.farmprogress.com/story-2013-wettest-spring-record-iowa-9-98723.

132 **"polluted runoff"** "Washout: Spring Storms Batter Poorly Protected Soil and Streams," Environmental Working Group, July 3, 2013, https://www.ewg.org/research/spring-storm-batter-midwest-soil-and-streams.

133 **"scars of erosion"** "Rain Damage Could Have Lasting Effect on Corn Crop," Agweb.com, June 20, 2013, https://www.agweb.com/article/rain_damage_could_have_lasting_effect_on_corn_crop.

133 **fourth-wettest year** Randy Schnepf, "Midwest Floods of 2008: Potential Impact on Agriculture," *Congressional Research Service*, August 18, 2008, https://nationalaglawcenter.org/wp-content/uploads/assets/crs/RL34583.pdf.

133 **without counting gulleys** Craig Cox, Andrew Hug, and Nils Bruzelius, "Losing Ground," Environmental Working Group, 2011, https://static.ewg.org/reports/2010/losingground/pdf/losingground_report.pdf.

133 **cataclysmic June floods** Federal Emergency Management Agency, "Midwest Floods of 2008 in Iowa and Wisconsin," July 26, 2013, https://www.fema.gov/media-library-data/20130726-1722-25045-4922/fema765.txt.

133 **20 tons per acre** AP, "Flooding, Heavy Rain Cause $40M in Soil Damage in Iowa," July 30, 2008, https://qctimes.com/news/state-and-regional/flooding-heavy-rain-cause-m-in-soil-damage-in-iowa/article_4adcaa45-c73c-5ff6-b450-63b1cabc67c4.html.

134 **water stress** Bernhard Schauberger et al, "Consistent Negative Response of US Crops to High Temperatures in Observations and Crop Models," *Nature Communications* 8 (2017): 13931, https://www.nature.com/articles/ncomms13931.

134 **hog herd stood at 10.7 million** USDA, Agricultural Census of 1950: Iowa, http://usda.mannlib.cornell.edu/usda/AgCensusImages/1950/01/09/1788/34102036v1p9ch1.pdf.

135 **every human Iowa resident** Chris Jones, "Iowa's Real Population," personal blog, March 14, 2019, https://www.iihr.uiowa.edu/cjones/iowas-real-population.

135 **56 million hens** "Industry Overview," American Egg Board, https://www.aeb.org/farmers-and-marketers/industry-overview.

135 **220,000 cows** USDA National Agricultural Statistics Service, "Milk Production," March 19, 2019, https://www.nass.usda.gov/Publications/Todays_Reports/reports/mkpr0319.pdf.

136 **55 "fecal equivalents"** "The 10 Largest Cities in the World," November 2, 2018, https://www.worldatlas.com/articles/the-10-largest-cities-in-the-world.html.

136 **highest fecal-equivalent density** Chris Jones, "50 Shades of Brown," personal blog, June 6, 2019, https://www.iihr.uiowa.edu/cjones/50-shades-of-brown.

136 **twelve thousand years ago** Amy Bogaard, Rebecca Fraser, Tim Heaton, et al., "Crop Manuring and Intensive Land Management by Europe's First Farmers," *Proceedings of the National Academy of Sciences* 110, no. 31 (2013): 12589–94.

136 **"processes of growth"** Albert Howard, *An Agricultural Testament* (London: Oxford University Press, 1940), 4.

137 **critique of industrial agriculture** Wendell Berry, *The Unsettling of America* (Berkeley, CA: Counterpoint Press, 2015 [1977]).

138 **lowered the price of nitrogen fertilizer** Gary Schnitkey, "Anhydrous Ammonia, Corn, and Natural Gas Prices over Time," *Farmdoc Daily* 6, June 14, 2016, https://farmdocdaily.illinois.edu/2016/06/anhydrous-ammonia-corn-and-natural-gas-prices.html.

138 **concentrate its toxins** "Toledo Issues Do-Not-Drink Advisory for Tap Water," NBC News, August 2, 2014, https://www.nbcnews.com/health/health-news/toledo-issues-do-not-drink-advisory-tap-water-n171236.

139 **a run on bottled water** "Tap Water Ban for Toledo Residents," *New York Times*, August 3, 2014, https://www.nytimes.com/2014/08/04/us/toledo-faces-second-day-of-water-ban.html.

139 **microcystin** Iowa Department of Health, "Fact Sheet: Microcystin Poisoning," https://www.idph.iowa.gov/Portals/1/Files/EHS/algae_faq.pdf.

139 **ozone-treatment system** Sarah Elms, "Big Renovations Continue at Toledo's Water Treatment Plant," *Toledo Blade*, May 8, 2019, http://www.toledoblade.com/local/city/2019/05/08/big-renovations-continue-at-toledo-collins-park-water-treatment-plant/stories/20190508014.

139 **lake's summer blooms** Ohio Department of Agriculture, "Ohio Lake Erie Phosphorus Task Force II Final Report 2013," https://lakeerie.ohio.gov/Portals/0/Reports/Task_Force_Report_October_2013.pdf.

140 **kind of algae** EPA, "Climate Change and Harmful Algal Blooms," *Nutrient Pollution*, March 9, 2017, https://www.epa.gov/nutrientpollution/climate-change-and-harmful-algal-blooms.

140 **more severe blooms** Douglas Smith, Kevin Kind and Mark Williams, "What Is Causing the Harmful Algal Blooms in Lake Erie?," *Journal of Soil and Water Conservation* 70, no. 2 (2015): 27A–29A, DOI: 10.2489/jswc.70.2.27A.

140 **increase in size and duration by mid-century** Steven Chapra, Brent Boehlert, Charles Fant, et al., "Climate Change Impacts on Harmful Algal Blooms in US Freshwaters: A Screening-Level Assessment," *Environmental Science and Technology* 51, no. 16 (2017): 8933–43, ttps://doi.org/10.1021/acs.est.7b01498.

140 **drained swamps and marshlands** Kyle Quandt, "Upper Maumee River Watershed Management Plan," February 5, 2012, https://www.in.gov/idem/nps/files/wmp_maumee-upper_wmp.pdf.

141 **more-efficient tile drains** Helen Jarvie, Laura Johnson, Andrew Sharpley, et al., "Increased Soluble Phosphorus Loads to Lake Erie: Unintended Consequences of Conservation Practices?," *Journal of Environmental Quality* 46, no. 1 (2017): 123–32, DOI: 10.2134/jeq2016.07.0248.

141 **leakage into the Maumee** "Interactive Map: Explosion of Unregulated Factory Farms," Environmental Working Group, https://www.ewg.org/interactive-maps/2019_maumee.

141 **restoring around 10 percent** William Mitsch, "Solving Lake Erie's Harmful Algal Blooms by Restoring the Great Black Swamp in Ohio," *Ecological Engineering* 108 (2017): 406–13, https://doi.org/10.1016/j.ecoleng.2017.08.040.

141 **drainage is viewed as sacred** Shannon Levy, "Learning to Love the Great Black Swamp," UnDark, March 31, 2017, https://undark.org/article/great-black-swamp-ohio-toledo.

141 **Lake Erie Bill of Rights** Amanda Paulson and Henry Gass, "Can a Lake Have Rights? Toledo Votes Yes," *Christian Science Monitor*, March 21, 2019,

https://www.csmonitor.com/Environment/2019/0321/Can-a-lake-have-rights-Toledo-votes-yes.

141 **"rights of nature"** Guillaume Chapron, Yaffa Epstein, and José Vicente López-Bao, "A Rights Revolution for Nature," *Science* 363, no. 6434 (2019): 1392–93, DOI: 10.1126/science.aav5601.

141 **Ohio Farm Bureau** Amanda Bush, "Farm Bureau to Support Farmer's Legal Action Against Lake Erie Bill of Rights," February 27, 2019, https://ofbf.org/2019/02/27/farm-bureau-support-farmers-legal-action-lake-erie-bill-of-rights.

142 **manure overapplication** "Trends in Swim Advisories," Iowa Environmental Council, https://www.iaenvironment.org/our-work/clean-water-and-land-stewardship/trends%20in%20beach%20advisories.

143 **natural gas refineries** Energy Information Administration, "Gulf of Mexico Fact Sheet," EIA, https://www.eia.gov/special/gulf_of_mexico.

144 **"Habitats that would normally be"** National Oceanic and Atmospheric Organization, "What Is a Dead Zone?," https://oceanservice.noaa.gov/facts/deadzone.html.

144 **blots out sea life** Rebecca Lindsey, "Wet Spring Linked to Forecast for Big Gulf of Mexico 'Dead Zone' This Summer," National Oceanic and Atmospheric Administration, June 18, 2019, https://www.climate.gov/news-features/features/wet-spring-linked-forecast-big-gulf-mexico-%E2%80%98dead-zone%E2%80%99-summer. State sizes from World Atlas, https://www.worldatlas.com/aatlas/infopage/usabysiz.htm.

144 **Iowa Farm Bureau informed** Preston Lyman, "Gulf of Mexico Dead Zone: What, Where, Why?," Iowa Farm Bureau, August 16, 2017, https://www.iowafarmbureau.com/Article/Gulf-of-Mexico-Dead-Zone-What-Where-Why.

145 **farms contribute 80 percent** USGS, "Sources of Nitrogen Delivered to the Gulf of Mexico," https://www.usgs.gov/media/images/sources-nitrogen-delivered-gulf-mexico-0.

145 **Iowa's nutrient-pollution** Christopher Jones, Jacob Nielsen, Keith Schilling, and Larry Weber, "Iowa Stream Nitrate and the Gulf of Mexico," *PLoS ONE* 13, no. 4 (2018): e0195930, https://doi.org/10.1371/journal.pone.0195930.

145 **blotting out life** Richard Manning, *Against the Grain: How Agriculture Has Hijacked Civilization* (New York: Macmillan, 2004).

145 **Gulf of Mexico hypoxia** Kevin Craig, "Hypoxia's Effects on the Shrimp Fishery in the Northwest Gulf of Mexico," National Centers for Coastal Ocean Science (2014), https://coastalscience.noaa.gov/project/hypoxias-effects-shrimp-fishery-northwest-gulf-mexico.

145 **less valuable shrimp** Martin Smith, Atle Oglend, A. Justin Kirkpatrick, et al., "Seafood Prices Reveal Impacts of a Major Ecological Disturbance," *Proceedings of the National Academy of Sciences* 114, no. 7 (2017): 1512–17, https://doi.org/10.1073/pnas.1617948114.

146 **bottom of the food chain** Kim De Mutsert, Jeroen Steenbeek, Kristy Lewis, et al., "Exploring Effects of Hypoxia on Fish and Fisheries in the Northern Gulf of Mexico Using a Dynamic Spatially Explicit Ecosystem Model," *Ecological Modelling* 331 (2016): 142–50, https://doi.org/10.1016/j.ecolmodel.2015.10.013.

146 **harms the reproductive capacity** C. Landry, S. L. Steele, S. Manning, and A. O. Cheek, "Long Term Hypoxia Suppresses Reproductive Capacity in the Estuarine Fish, *Fundulus grandis*," *Comparative Biochemistry and Physiology Part A: Molecular and Integrative Physiology* 148, no. 2 (2007): 317–23. See also Peter Thomas and M. Saydur Rahman, "Extensive Reproductive Disruption, Ovarian Masculinization and Aromatase Suppression in Atlantic Croaker in the Northern Gulf of Mexico Hypoxic Zone," *Proceedings of the Royal Society B: Biological Sciences* 279, no. 1726 (2011): 28–38.

146 **emitters of nitrous oxide** Louis Codispoti, "Interesting Times for Marine N_2O," *Science* 327, no. 5971 (2010): 1339–40, DOI: 10.1126/science.1184945.

146 **accelerated by climate change** Il-Nam Kim, "Estimating 'Mean-State' July (1985–2007) N2O Fluxes in the Northern Gulf of Mexico Hypoxic Region: Variation, Distribution, and Implication," *Frontiers in Marine Science* 5 (2018): 249, https://doi.org/10.3389/fmars.2018.00249.

147 **slammed U.S. corn yields** "Crop Production Down in 2012 Due to Drought, USDA Reports," USDA NASS, January 11, 2013 https://www.nass.usda.gov/Newsroom/archive/2013/01_11_2013.php.

147 **losses reverberated** Karl Plume and Deborah Zabarenko, "Grain Prices Set Records as U.S. Drought, Food Worries Spread," Reuters, July 19, 2012, https://www.reuters.com/article/us-usa-drought/grain-prices-set-records-as-drought-food-worries-spread-idUSBRE86F1D420120720.

CHAPTER 7: THE BIG LIFT

151 **covered just 3 percent of Iowa's farm acres** Liz Juchems, "Cover Crop Acres Increase but Rate of Growth Declines in 2018," Iowa State University Extension, March 15, 2019, https://www.extension.iastate.edu/news/cover-crop-acres-increase-rate-growth-declines-2018.

152 **U.S. bird population had plunged by nearly 30 percent** Kenneth Rosenberg, Adriaan Dokter, Peter Blancher, et al., "Decline of the North American Avifauna," *Science*, September 19, 2019, DOI: 10.1126/science.aawl313.

153 **between 1966 and 2013** C. B. Wilsey, J. Grand, J. Wu, et al., "North American Grasslands and Birds Report" *National Audubon Society*, 2019, https://www.audubon.org/sites/default/files/audubon_north_american_grasslands_birds_report-final.pdf.

153 **variety of hybrid rye** Molly McGhee and Hans Stein, "Hybrid Rye Holds Promise as Feed Ingredient in North America," August 30, 2018, https://www.nationalhogfarmer.com/nutrition/hybrid-rye-holds-promise-feed-ingredient-north-america.

154 **grown on pastures** Chris Mayda, "Pig Pens, Hog Houses, and Manure Pits: A Century of Change in Hog Production," *Material Culture* 36, no. 1 (2004): 18–42, https://www.jstor.org/stable/29764207.

154 **three main crops grown in rotation** For corn, soybeans, and oats: "The Decline of Oat Production in Iowa Fields," *Decision Innovation*, July 20, 2016, http://www.decision-innovation.com/blog/disinsights/the-decline-of-oat-production-in-iowa-fields; for hay: "Hay—Living History Farms, Iowa" *Learning Fields*, https://www.lhf.org/learning-fields/crops/hay.

155 **three times more water flowing from farms into streams** Chris Jones, "This Is What Happened," *IIHR Research Engineer*, February 12, 2019, https://www.iihr.uiowa.edu/cjones/this-is-what-happened.

156 **hay crops they displaced** Scott Killpack and Daryl Buchholz, "Nitrogen in the Environment: Nitrogen Replacement Value of Legumes," *Water Quality Initiative Publication*, 1993, https://extension2.missouri.edu/wq277.

156 **160 pounds** Jones, "This Is What Happened."

156 **Rates for phosphorus** "Fertilizer Use and Price," USDA ERS, https://www.ers.usda.gov/data-products/fertilizer-use-and-price.aspx.

156 **nitrogen fertilization rates** Jim Camberato, "A Historical Perspective on Nitrogen Fertilizer Rate Recommendations for Corn in Indiana (1953–2011)," *Purdue Extension*, 2015, https://extension.purdue.edu/extmedia/ay/ay-335-w.pdf.

156 **denying the water-pollution crisis** Craig Hill, "Iowa Farmers Make True Progress on Water Quality," *Iowa Farm Bureau*, August 1, 2019, https://www.iowafarmbureau.com/Article/Iowa-farmers-make-true-progress-on-water-quality.

156 **optimal crop yields** Darcy Maulsby, "Working Together to Improve Water Quality," *Iowa Farm Bureau*, May 6, 2019, https://www.iowafarmbureau.com/Article/Working-together-to-improve-water-quality.

157 **corn-soy duopoly** Adam Davis, Jason Hill, Craig Chase, Ann Johanns, and Matt Liebman, "Increasing Cropping System Diversity Balances Productivity, Profitability and Environmental Health," *PLoS ONE* 7:10 (2012): e47149. https://doi.org/10.1371/journal.pone.0047149. See also Natalie Hunt, Jason Hill, and Matt Liebman, "Cropping System Diversity Effects on Nutrient Discharge, Soil Erosion, and Agronomic Performance," *Environmental Science and Technology* 53, no. 3 (2019): 1344–52, https://doi.org/10.1021/acs.est.8b02193.

158 **22 percent beneath normal** Gary Schnitkey, "Locating the 2012 Drought Using State Corn Yields," *Farmdoc Daily* 3 (2013), https://farmdocdaily.illinois.edu/2013/02/locating-the-2012-drought.html.

160 **carbon sequestration** D. C. Reicosky, "Tillage-Induced CO_2 Emission from Soil," *Nutrient Cycling in Agroecosystems* 49, no. 1–3 (1997): 273–85, https://link.springer.com/article/10.1023/A:1009766510274.

CHAPTER 8: THE FUTURE OF THE FARM

166 **massive snowpack** California Department of Water Resources, "Snow Survey Boosts Runoff Predictions," April 2, 2019, https://water.ca.gov/News/News-Releases/2019/April/Snow-Survey-Boosts-Runoff-Predictions.

166 **Water anxiety** Dale Kasler, "California Farmers Fear 'Catastrophe' from Water Restrictions," *Sacramento Bee*, September 25, 2019, https://www.sacbee.com/news/california/big-valley/article233596597.html.

167 **half the broccoli** Rosanna Xia, "Salinas Valley's Thriving Crops Mask Fears over the Area's Lone Water Source," *Los Angeles Times*, September 7,

2015, https://www.latimes.com/local/california/la-me-drought-salinas-valley-20150907-story.html. See also "A Look at Year-Round Lettuce Production—from California's Leafy Green's Marketing Agreement," California Ag Network, November 3, 2017, https://californiaagnet.com/2017/11/03/a-look-at-year-round-lettuce-production-from-californias-leafy-greens-marketing-agreement; and "A Look at Year-Round Lettuce Production," California Ag Network, November 3, 2017, https://californiaagnet.com/2017/11/03/a-look-at-year-round-lettuce-production-from-californias-leafy-greens-marketing-agreement.

168 **pushing up prices** Aaron Kinney, "Boom Time for Farmers? Salinas Valley Thrives, for Now," *Mercury News*, July 7, 2015, https://www.mercurynews.com/2015/07/07/boom-time-for-farmers-salinas-valley-thrives-for-now.

168 **nitrate levels over the legal limit** Kenneth Belitz, Miranda Fram, and Tyler Johnson, "Metrics for Assessing the Quality of Groundwater Used for Public Supply, CA, USA: Equivalent-Population and Area," *Environmental Science and Technology* 49, no. 14 (2015): 8330–38, https://doi.org/10.1021/acs.est.5b00265.

168 **ninety thousand farmworkers** Samantha Michaels, "Farmworkers Are Living 20 to a House in California's Bountiful Salinas Valley," *Mother Jones*, April 21, 2018, https://www.motherjones.com/food/2018/04/farmworkers-are-living-20-to-a-house-in-californias-bountiful-salinas-valley.

168 **poverty line** Alex Vandermaas-Peeler, Daniel Cox, Maxine Najle, Molly Fisch-Friedman, Rob Griffin, and Robert P. Jones, "A Renewed Struggle for the American Dream," PRRI, August 28, 2018, https://www.prri.org/research/renewed_struggle_for_the_american_dream-prri_2018_california_workers_survey.

168 **moratorium on digging new wells** Monterey County Water Resources Agency, "Recommendations to Address Seawater Intrusion in the Salinas County Groundwater Basin," October 2017, https://www.co.monterey.ca.us/home/showdocument?id=57394.

168 **limit their future water access** Jim Johnson, "Salinas Valley Wells Moratorium Gets Thumbs Up over Ag Concerns," *Monterey Herald*, April 24, 2018, https://www.montereyherald.com/2018/04/24/salinas-valley-wells-moratorium-gets-thumbs-up-over-ag-concerns.

168 **pricier salad greens** Jim Johnson, "Salinas Valley Basin Draft Plan Proposes Millions in Projects," *Monterey Herald*, September 6, 2019, https://www

.montereyherald.com/2019/09/06/salinas-valley-basin-draft-plan-proposes-millions-in-projects-management-actions.

168 **Then there's the Imperial Valley** Imperial County Farm Bureau, "Quick Facts About Imperial County Agriculture," https://www.co.imperial.ca.us/AirPollution/Forms%20&%20Documents/AGRICULTURE/QuickFactsAboutIVag.pdf.

169 **14 percent of all leafy greens** "A Look at Year-Round Lettuce Production," California Ag Network, November 3, 2017, https://californiaagnet.com/2017/11/03/a-look-at-year-round-lettuce-production-from-californias-leafy-greens-marketing-agreement.

169 **13.4 million acre-feet** U.S. Department of the Interior, "Drought in the Colorado River Basin," https://www.doi.gov/water/owdi.cr.drought.

169 **Rocky Mountain snowpack** Amato Evan, "A New Method to Characterize Changes in the Seasonal Cycle of Snowpack," *Journal of Applied Meteorology and Climatology* 58, no. 1 (2019): 131–43, https://doi.org/10.1175/JAMC-D-18-0150.1.

169 **dwindle by another 30 percent by midcentury** Bradley Udall and Jonathan Overpeck, "The Twenty-First Century Colorado River Hot Drought and Implications for the Future," *Water Resources Researc*h 53, no. 3 (2017): 2404–18, https://agupubs.onlinelibrary.wiley.com/doi/full/10.1002/2016WR019638.

170 **Salton Sea, the original sin** Eric Niiler, "Salton Sea Resort Hoping for Return of Glory Days," *San Diego Union-Tribune*, June 30, 1998, http://www.sci.sdsu.edu/salton/NiilerSaltonSeaResort.html.

171 **asthma hospitalizations** California Department of Public Health, "Imperial County Asthma Profile," September 1, 2016, https://www.cdph.ca.gov/Programs/CCDPHP/DEODC/EHIB/CPE/CDPH%20Document%20Library/County%20profiles/Imperial2016profile.pdf.

171 **hydrogen sulfide** Ian Lovett, "Lake Is Blamed for Stench Blown Across Southern California," *New York Times*, September 11, 2012, https://www.nytimes.com/2012/09/12/us/salton-sea-is-blamed-for-southern-california-stench.html.

171 **Colorado Basin lost** Stephanie L. Castle, Brian F. Thomas, John T. Reager, Matthew Rodell, Sean C. Swenson, and James S. Famiglietti, "Groundwater Depletion During Drought Threatens Future Water Security of the Colorado River Basin," *Geophysical Research Letters* 41, no. 16 (2014): 5904–11.

171 **"we are alarmed"** James S. Famiglietti, "How the West Was Lost," *National Geographic*, July 24, 2014, http://blog.nationalgeographic.org2014/07/24/how-the-west-was-lost.

172 **Saudi wheat production** Javier Blas, "Saudi Wells Running Dry—of Water—Spell End of Desert Wheat," *Bloomberg*, November 3, 2015, https://www.bloomberg.com/news/articles/2015-11-04/saudi-wells-running-dry-of-water-spell-end-of-desert-wheat.

172 **town of Tayma** Nathan Halverson, "What California Can Learn from Saudi Arabia's Water Mystery," *Reveal News*, April 22, 2015, https://www.revealnews.org/article/what-california-can-learn-from-saudi-arabias-water-mystery.

173 **make-the-desert-bloom ambitions** Yossi Mann, "Can Saudi Arabia Feed Its People?," *Middle East Quarterly*, Spring 2015, https://www.meforum.org/5098/can-saudi-arabia-feed-its-people.

173 **importing substantial amounts of alfalfa** Daniel H. Putnam, Bill Matthews, and Daniel A. Sumner, "Growth in Saudi Hay Demand Likely to Have World-Wide Impacts," *Alfalfa & Forage News*, March 7, 2017, https://aic.ucdavis.edu/2017/03/08/growth-in-saudi-hay-demand-likely-to-have-world-wide-impacts/

173 **source of 44 percent of U.S. fruit** Renée Johnson, "The US Trade Situation for Fruit and Vegetable Products," *Congressional Research Service*, 2016.

174 **lack running water** Felicity Barringer, "Strawberry Fields Forever? Thirsty Baja Turning to Seawater to Grow Lucrative Crop," *And the West Blog*, Stanford University, July 13, 2018, https://west.stanford.edu/news/blogs/and-the-west-blog/2018/strawberry-fields-forever-baja-eyes-desalination-grow-lucrative-crop.

174 **Mexico's water balance** "Aqueduct Country Ranking," World Resources Institute, https://www.wri.org/applications/aqueduct/country-rankings/?country=MEX.

174 **Chile, a heavy contributor** René Garreaud, Juan P. Boisier, Roberto Rondanelli, Aldo Montecinos, Hector H. Sepúlveda, and Daniel Veloso-Aguila, "The Central Chile Mega Drought (2010–2018): A Climate Dynamics Perspective," *International Journal of Climatology*, 2019, https://doi.org/10.1002/joc.6219.

174 **wells have gone dry** Nicky Milne, "As Sales Boom, Chile's 'Green Gold' Is Blamed for Water Shortages," Reuters, June 3, 2019, https://www.reuters.com/article/us-water-chile-environment/as-sales-boom-chiles-green-gold-is-blamed-for-water-shortages-idUSKCN1T41AL.

174 **imports from Chile** Nick Austin, "U.S. Shipments from Chile Dropping Amid Extreme Drought," *Freight Waves*, August 19, 2019, https://www

.freightwaves.com/news/u-s-shipments-from-chile-dropping-amid-extreme-drought.

174 **large importer of produce** California Department of Food and Ag, "Agricultural Statistics Review, 2017–18," https://www.cdfa.ca.gov/statistics/PDFs/2017-18AgReport.pdf.

176 **U.S. farmers markets** USDA Agricultural Marketing Service, "Number of Operating Farmers Markets," https://www.ams.usda.gov/sites/default/files/media/NationalCountofFarmersMarketDirectoryListings082019.pdf.

176 **until the early 1980s** Steven McFadden, "Community Farms in the 21st Century: Poised for Another Wave of Growth?," Rodale Institute, http://www.newfarm.org/features/0104/csa-history/part1.shtml.

176 **$226 million in sales** USDA National Agriculture Library, "Community Supported Agriculture," https://www.nal.usda.gov/afsic/community-supported-agriculture.

176 **farm sales directly to consumers** USDA National Agricultural Statistics Service, "Direct Farm Sales of Food," 2016, https://www.nass.usda.gov/Publications/Highlights/2016/LocalFoodsMarketingPractices_Highlights.pdf.

177 **boasts 220 farmers markets** Iowa Farm Bureau, "Iowans Love Farmers Markets," July 29, 2018, https://www.iowafarmbureau.com/Article/Iowans-love-farmers-markets.

177 **federal food-aid programs** USDA Economic Research Service, "Trends in US Local and Regional Food Systems," January 2015, https://www.ers.usda.gov/webdocs/publications/42805/51173_ap068.pdf.

177 **midsize farms** Fred Kirschenmann, G. W. Stevenson, Frederick Buttel, Thomas Lyson, and Mike Duffy, "Why Worry About the Agriculture of the Middle?," in *Food and the Mid-Level Farm: Renewing an Agriculture of the Middle*, eds. Thomas Lyson et al. (Cambridge: MIT Press, 2008), 3–22.

179 **Iowa farmers** Rich Pirog, Timothy Van Pelt, Kamyar Enshayan, and Ellen Cook, "Food, Fuel, and Freeways: An Iowa Perspective on How Far Food Travels, Fuel Usage, and Greenhouse Gas Emissions," Leopold Center Publications and Papers, 2001, https://lib.dr.iastate.edu/leopold_pubspapers/3.

179 **vegetables trucked into a region** Ken Meter and Jon Rosales, "Finding Food in Farm Country: The Economics of Food and Farming in Southeast Minnesota," Crossroads Resource Center, 2001, https://www.crcworks.org/ff.pdf.

180 **corn and soybean production** Dave Swenson, "Selected Measures of the Economic Values of Increased Fruit and Vegetable Production and Consumption in the Upper Midwest," Leopold Center Publications and Papers, 2010, https://www.leopold.iastate.edu/files/pubs-and-papers/2010-03-selected-measures-economic-values-increased-fruit-and-vegetable-production-and-consumption-upper-mid.pdf.

182 **$9 million annually** "Senate Agriculture, Nutrition and Forestry Committee," Open Secrets, https://www.opensecrets.org/cong-cmtes/overview?cmte=SAGR&cmtename=Senate+Agriculture%2C+Nutr+%26+Forestry+Committee&cong=115.

182 **$8 million** "House Agriculture Committee Profile," Open Secrets, https://www.opensecrets.org/cong-cmtes/overview?cmte=HAGR&cmtename=House+Agriculture+Committee&cong=109.

182 **around $100 million** "Totals by Sector," Open Secrets, https://www.opensecrets.org/overview/sectors.php.

182 **$2.5 billion** "Totals by Sector," Open Secrets, https://www.opensecrets.org/federal-lobbying/ranked-sectors?cycle=a.

182 **construction firms** "Lobbying Spending Database," Open Secrets, https://www.opensecrets.org/lobby/top.php?indexType=c.

182 **Monsanto and Bayer agreed to merge** Soo Rin Kim, "Bayer-Monsanto Merger: Two Washington-Savvy Companies Get Their Game On," Open Secrets, September 15, 2016, https://www.opensecrets.org/news/2016/09/bayer-monsanto-merger-two-washington-savvy-companies-get-their-game-on.

183 **Bayer continues bombarding** "Lobbying Spending Database—Bayer AG," Open Secrets, https://www.opensecrets.org/lobby/clientsum.php?id=D000042363.

183 **chief scientist at the US Department of Agriculture** "President Donald J. Trump Announces Intent to Nominate Personnel," White House, https://www.whitehouse.gov/presidential-actions/president-donald-j-trump-announces-intent-nominate-personnel-key-administration-post-9.

183 **$2.9 billion research budget** "FY 2020 United States Department of Agriculture Budget Summary," USDA, https://www.obpa.usda.gov/budsum/fy2020budsum.pdf.

183 **USDA's foreign-trade agency** USDA, "Secretary Perdue Announces New Senior Leaders at USDA," April 19, 2018, https://www.usda.gov/media/press-releases/2018/04/19/secretary-perdue-announces-new-senior-leaders-usda.

183 **pushing for pesticide deregulation** "Client Profile: Corteva Agriscience," Open Secrets, 2019, https://www.opensecrets.org/federal-lobbying/clients/summary?id=D000019355.

184 **costs to farmers** Richard Howitt, Josué Medellín-Azuara, Duncan MacEwan, Jay Lund, and Daniel Sumner, "Economic Analysis of the 2014 Drought for California Agriculture," Center for Watershed Sciences, University of California at Davis, July 23, 2014, https://watershed.ucdavis.edu/files/biblio/DroughtReport_23July2014_0.pdf.

184 **food prices barely budged** Codi Kozacek, "California Drought Effect on Food Prices—Not Much," Circle of Blue, May 26, 2015, https://www.circleofblue.org/2015/world/california-drought-effect-on-food-prices-not-much.

184 **ethanol sucked** Energy Policy Act of 2005, Congress.gov, August 8, 2005, https://www.congress.gov/109/plaws/publ58/PLAW-109publ58.pdf.

185 **bumper harvests** "U.S. Corn Used for Ethanol," Ag Update, April 24, 2018, https://www.agupdate.com/u-s-corn-used-for-ethanol/pdf_bbefcecc-5c55-54c3-82c2-7a42325fabda.html. See also Department of Energy, "Maps and Data: Alternative Fuels Data Center," https://afdc.energy.gov/data.

185 **soybean yields to plunge** "Corn Price—45 Year Historical Chart," Macro Trends, https://www.macrotrends.net/2532/corn-prices-historical-chart-data.

185 **corn prices spiked** "Corn Belt Disaster in Wake of Record Heat Wave," AccuWeather, July 19, 2012, accessed July 25, 2019, https://www.accuweather.com/en/weather-news/record-heat-wave-resulting-in/67651.

185 **global food crisis** John Vidal, "UN Warns of Looming Worldwide Food Crisis in 2013," *The Guardian*, October 13, 2012, https://www.theguardian.com/global-development/2012/oct/14/un-global-food-crisis-warning.

185 **corn as a staple** Emily Nuss and Sherry Tanumihardjo, "Maize: A Paramount Staple Crop in the Context of Global Nutrition," *Comprehensive Reviews in Food Science and Food Safety* 9, no. 4 (2010): 417–36, https://onlinelibrary.wiley.com/doi/full/10.1111/j.1541-4337.2010.00117.x.

185 **"tortilla riots"** Adam Thomson, "'Tortilla Riots' Give Foretaste of Food Challenge," *Financial Times*, October 12, 2010, https://www.ft.com/content/a0aa9ef0-d618-11df-81f0-00144feabdc0.

185 **44 million people** "The Global Food Crises," United Nations, https://www.un.org/esa/socdev/rwss/docs/2011/chapter4.pdf.

186 **box of cornflakes** Ephraim Leibtag, "Corn Prices Near Record High, But What About Food Costs?," USDA Economic Research Service, February 1, 2008, https://www.ers.usda.gov/amber-waves/2008/february/corn-prices-near-record-high-but-what-about-food-costs.

187 **have to fallow 780,000** Ellen Hanak, Alvar Escriva-Bou, Brian Gray, et al., "Water and the Future of the San Joaquin Valley," Public Policy Institute of California, 2019, https://www.ppic.org/publication/water-and-the-future-of-the-san-joaquin-valley.

187 **Union Square Greenmarket** Michael Pollan, "Voting with Your Fork," *New York Times*, May 7, 2006, https://michaelpollan.com/articles-archive/voting-with-your-fork.

189 **atmospheric CO_2** Daily CO_2, CO_2.Earth, https://www.co2.earth/daily-co2#.

INDEX

Note: page numbers in italics refer to figures.

A NOTE ON THE AUTHOR

TOM PHILPOTT has been the food and agriculture correspondent for *Mother Jones* since 2011. Previously, he covered food as a writer and editor for the environmental-news website *Grist*. Philpott's work on food politics has won numerous awards, including the Gerald Loeb Award for business feature writing, and has appeared in the *New York Times*, *Newsweek*, and the *Guardian*, among other places. From 2004 to 2012, he farmed at Maverick Farms in Valle Crucis, North Carolina. He lives in Austin, Texas.